Protein Chemistry

Protein Chemistry

Edited by Nigel Stokes

SYRAWOOD
PUBLISHING HOUSE

New York

Published by Syrawood Publishing House,
750 Third Avenue, 9th Floor,
New York, NY 10017, USA
www.syrawoodpublishinghouse.com

Protein Chemistry
Edited by Nigel Stokes

International Standard Book Number: 978-1-68286-596-5 (Hardback)

Cataloging-in-Publication Data

Protein chemistry / edited by Nigel Stokes.
    p. cm.
Includes bibliographical references and index.
ISBN 978-1-68286-596-5
1. Proteins. 2. Proteins--Analysis. 3. Biotechnology. I. Stokes, Nigel.
QD431 .P76 2018
572.6--dc23

TABLE OF CONTENTS

PREFACE

Proteins are organic compounds which are formed of amino acids that are linked together by peptides. They help the body in getting nitrogen, vitamins and sulfur. Proteins are three dimensional in their structure. Their structure can be categorized into four distinctive aspects - primary structure, secondary structure, quaternary structure and tertiary structure. As this subject is emerging at a rapid pace, the contents of this book will help the readers understand the modern concepts and applications of the subject. This book is meant for students who are looking for an elaborate reference text on protein chemistry.

To facilitate a deeper understanding of the contents of this book a short introduction of every chapter is written below:

Chapter 1- To develop a better understanding of how protein develops its structure, it is very important to understand the different levels of protein. The four structures of protein are primary, secondary, tertiary and quaternary. This chapter has been carefully written to provide an easy understanding of the varied structures of protein.

Chapter 2- Proteins change their color or structure when some external pressure is applied on them. They must be folded in the correct shape for them to function properly. The common example cited in the denaturing of proteins is that of egg whites. The topics discussed in the chapter are of great importance to broaden the existing knowledge on protein chemistry.

Chapter 3- Proteins are of various types with a vast arrangement of functions within an organism. Some of the types of proteins are metalloproteins, motor proteins, membrane proteins, etc. Mostly proteins belong to the category of metalloproteins. The major categories of protein are dealt with great details in this chapter.

Chapter 4- Electron transfer helps in the rearrangement of atoms from one chemical entity to another. Several processes, such as photosynthesis and respiration, which are very important to an organism involve electron transfer. The three main types of electron transfer are heterogeneous electron transfer, inner-sphere electron transfer and outer-space electron transfer. The diverse functions of electrons and enzymes have been thoroughly discussed in this chapter.

I owe the completion of this book to the never-ending support of my family, who supported me throughout the project.

Editor

Structure of Protein

To develop a better understanding of how protein develops its structure, it is very important to understand the different levels of protein. The four structures of protein are primary, secondary, tertiary and quaternary. This chapter has been carefully written to provide an easy understanding of the varied structures of protein.

Protein

Proteins are the most abundant organic molecules in living cells. They may be monomeric with one polypeptide chain or multimeric having more than one chain.

In case of a homomultimer the chains are of one kind whereas for a heteromultimer two or more different chains form the protein. (e.g. Hemoglobin is a heterotetramer. It has two alpha chains and two beta chains).

Proteins may be simple or conjugated – Simple – composed only of amino acid residues

Conjugated – in addition to the polypeptide chain these proteins contain other non-amino acid components known as prosthetic groups (e.g. metal ions, cofactors, lipids, carbohydrates) Example: Hemoglobin – Heme

Each polypeptide chain which is a polymer of amino acids linked by peptide bonds can be classified according to their shape and/or function.

Some common terms:

- Peptide = a short chain of amino acids
- Polypeptide = a longer chain of amino acids
- Protein = a polypeptide that occurs in nature and folds into a defined three-dimensional structure

This Means that ALL Proteins are Polypeptides but not all Polypeptides are Proteins.

The only other covalent bonds present in proteins apart from the peptide bonds (or any post translational modification that may from covalent bonds) are disulfide bonds.

The disulfide bonds are able to link parts of the polypeptide chain that may otherwise be separated by a large number of amino acid residues.

Protein Structure

- Physical properties of proteins that influence stability
  - Rigidity of backbone
  - Amino acid interactions with water
  - Interactions among amino acids
- Electrostatic interactions
- Hydrogen bonds
- S-S bonds
- Volume constraints

Protein Hierarchy

Primary structure (1°) : the amino acid sequence.

Secondary structure (2°) : helices, sheets and turns.

Tertiary structure (3°) : side chain packing in the 3-D structure.

Quaternary structure (4°) : association of subunits.

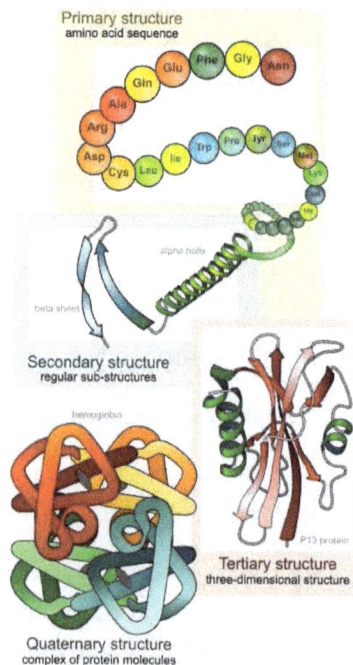

Primary structure
amino acid sequence

Secondary structure
regular sub-structures

Tertiary structure
three-dimensional structure

Quaternary structure
complex of protein molecules

The primary structure is the sequence of amino acid side chains along the backbone of peptide bonds in a polypeptide. Considering the 20 common amino acids only, with no limitation as to the length or the type of amino acid to be linked, there are a huge number of possibilities for the primary structure.

For example if we would want to determine how many primary structures with three amino acids are possible, we would see that there are 8000 possibilities.

There are 20 possibilities for the first place in the chain. Similarly there are 20 possibilities for the second place in the chain, making a total of $20 \times 20 = 400$ possible combinations for a dipeptide. Extending this to a tripeptide means there are $400 \times 20 = 8000$ possibilities!

With the polypeptide chains ranging from 20-1000+ residues the possibilities are immense.

The primary structure of a protein is conventionally specified by writing the three-letter abbreviations for each amino acid. The amino terminal is the $-NH_3^+$ end of the polypeptide chain. This is the convention because protein synthesis starts from the N-terminal end.

In some cases, this is even shortened to the 1-letter abbreviation sequence.

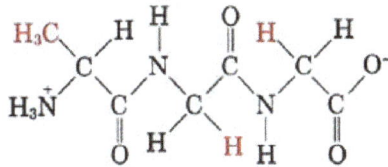

For example, the structure which reading from the $-NH_3^+$ end is alanine, glycine, glycine would be specified as Ala-Gly- Gly in the three-letter code format and simply as AGG in the one letter code format.

The primary sequence of a protein determines the three-dimensional shape that ultimately relates to the function of the protein. Therefore a lot of effort has gone into determining the primary sequence of proteins. The common method for amino acid sequence determination is called Edman degradation.

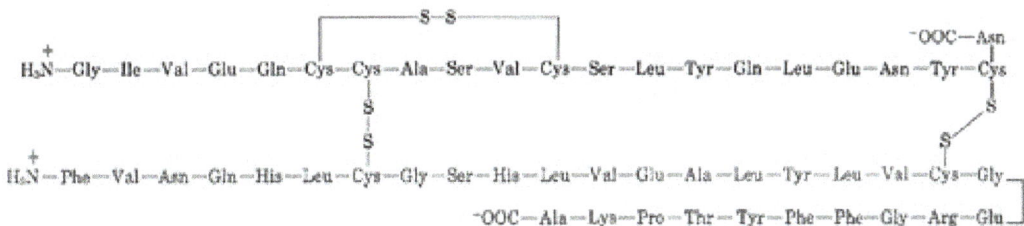

The first protein whose amino acid sequence was determined is insulin. This was determined by Frederick Sanger in 1953 after many years of effort. There are two chains in this structure that are linked together by disulfide bonds between Cys residues of the different chains.

The secondary structure involves a specific geometry of the polypeptide backbone where the backbone atoms are linked by hydrogen bonds. The defined backbone angles depicted by the φ and ψ angles result in a regular pattern of the polypeptide chain.

We have seen φ and ψ angles and realize that rotation about these bonds can change the corresponding dihedral angles of the backbone. This brings about conformational flexibility to the molecule and allows distant regions of the polypeptide chain to come close to one another.

As we have seen earlier the other possibility that may bring distant part of the polypeptide close to one another is the disulfide bond.

The question now is – do the backbone angles have any regularity or are they completely random? To answer this question, we need to understand a bit more of the geometry or conformation of protein structures.

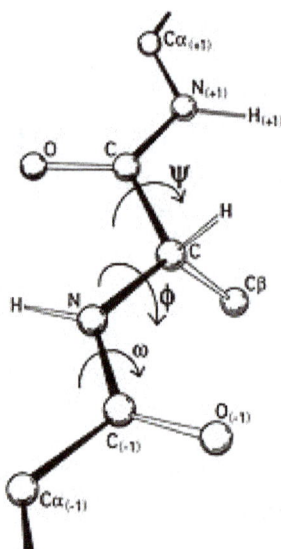

The α-helix

The α-helix is one of two secondary structures (the other being the β -sheet) predicted and discovered by Linus Pauling in 1951.

Helix Nomenclature:

1.   Hydrogen bond (dotted line) between

         >C=O-------------H-N<

     (residue i)   (residue i+3/4/5)

Repeating unit:

(i) 3_{10} helix

   - 10 atoms

   - 3 residues per turn

   - i to i+3 hydrogen bonding

(ii) 3.6_{13} helix

   - 13 atoms

   - 3.6 residues per turn

   - i to i+4 hydrogen bonding

(iii) 5_{18} helix

   - 18 atoms

   - 5 residues per turn

   - i to i+5 hydrogen bonding

A regular right-handed α-helix has the following spatial parameters:

$\phi = -57°$

$\psi = -47°$

n =3.6 (number of residues per turn) pitch 0.54nm (or 5.4Å)

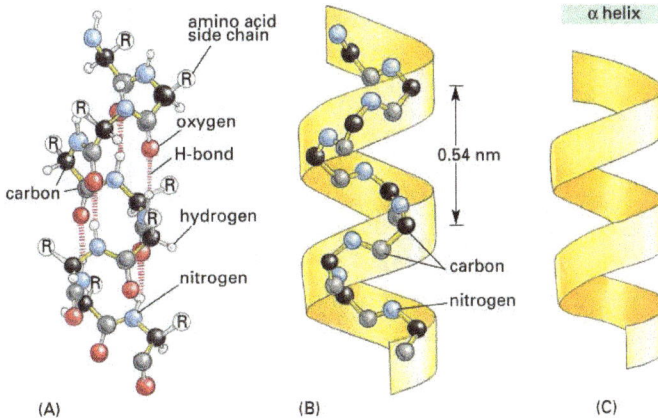

(A)   (B)   (C)

The β-sheet

The β-sheet forms a result of H-bonding between polypeptide chains that are referred to as strands in this case.

β-sheet: parallel (φ = -119 and ψ =113) or anti-parallel pleated sheet structures

A four-stranded beta sheet that contains three antiparallel strands and one parallel strand is shown. Hydrogen bonds are indicated with red lines (antiparallel strands) and blue lines (parallel strands). Note the difference in the hydrogen bonding pattern.

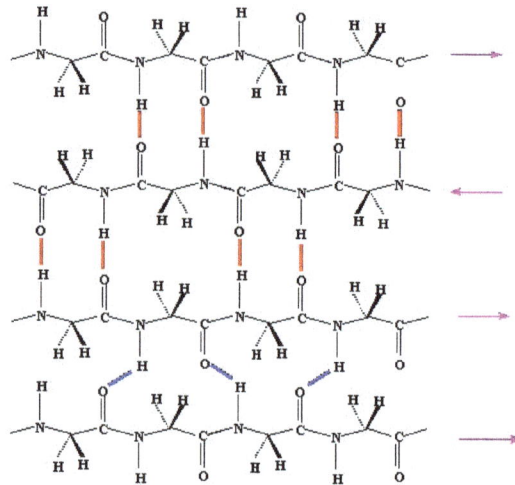

A Ramachandran plot, often referred to as a φ,ψ plot was developed in 1963 by G. N. Ramachandran and colleagues to be able to understand the geometrical features of the backbone. It was demonstrated that the backbone dihedral angles ψ against φ of amino acid residues in the protein structure are unique for specific secondary structural elements as shown.

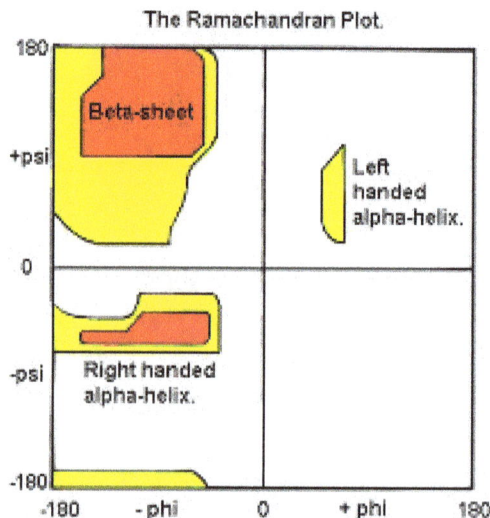

The plot is usually used to determine which conformations are possible for a protein but its usage is very important in structure validation of proteins. Since some regions of the map are forbidden due to geometrical restraints in the backbone, the calculations

of the dihedral angles serve as a measure of the structural integrity of the protein structure.

Turns – short regions of non-α, non-β conformation. These are also referred to the coil regions of the protein and they act as linkers of the secondary structural elements the α-helix and β-sheet.

This leads to the formation of a tertiary structure that is held together primarily by noncovalent interactions which we will see later.

The different levels of structure are given below:

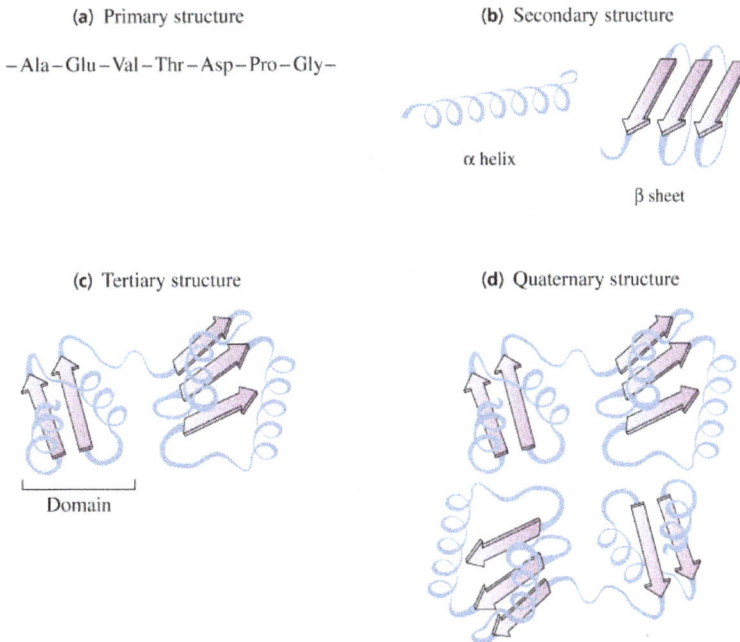

(a) Primary structure

– Ala – Glu – Val – Thr – Asp – Pro – Gly –

(b) Secondary structure

α helix

β sheet

(c) Tertiary structure

Domain

(d) Quaternary structure

Principle Forces in Protein Folding

- van der Waals interactions along with hydrophobic forces that occur favors packing of the amino acid interactions and are responsible for the formation of a hydrophobic core of the protein. This exploits the hydrophobic tendency of non-polar amino-acid side-chains.

- Polar side-chain interactions occur either through hydrogen bonds or electrostatic interactions.

- Disulfide bond formation: provides stability

Structure Stabilizing Interactions in secondary structure of proteins are primarily hydrogen bonding as seen earlier. This has a higher interaction energy than a simple van der Waals interaction

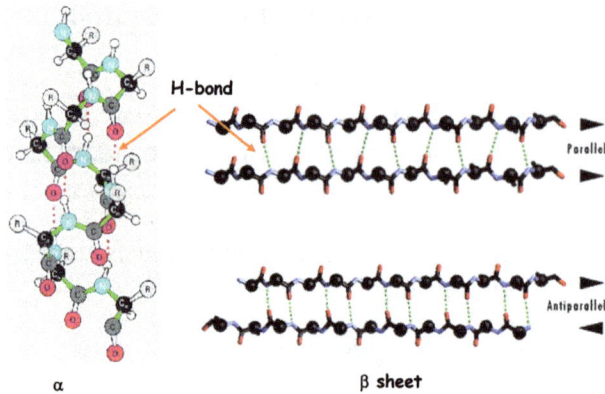

H-bond

Parallel

Antiparallel

α   β sheet

Overall interactions that comprise the tertiary structure of proteins

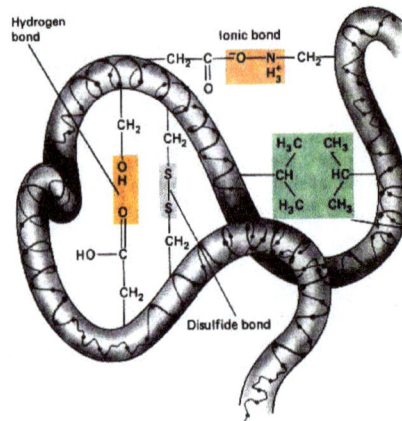

Hydrogen bond

Ionic bond

Disulfide bond

Every protein has at least three levels of structural organization. Some of them may have a fourth level giving rise to what is known as the quaternary structure of proteins where monomeric subunits interact to form a multimeric protein. The most common example that can be thought of is hemoglobin which is a tetrameric protein having two α-subunits and two β-subunits. These assemble to form the tetrameric structure of hemoglobin.

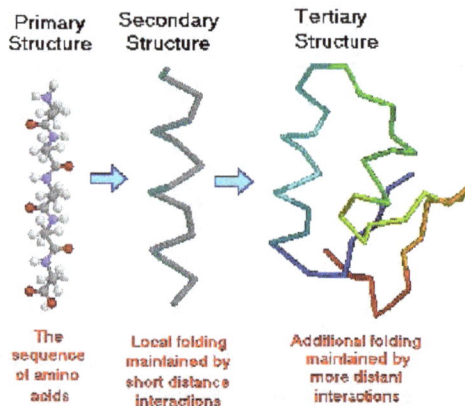

Primary Structure	Secondary Structure	Tertiary Structure
The sequence of amino acids	Local folding maintained by short distance interactions	Additional folding maintained by more distant interactions

Structure of human hemoglobin: The α and β subunits are in red and blue, and the iron- containing heme groups in green. So hemoglobin is actually a dimer of dimers and can be referred to as a heteromeric protein.

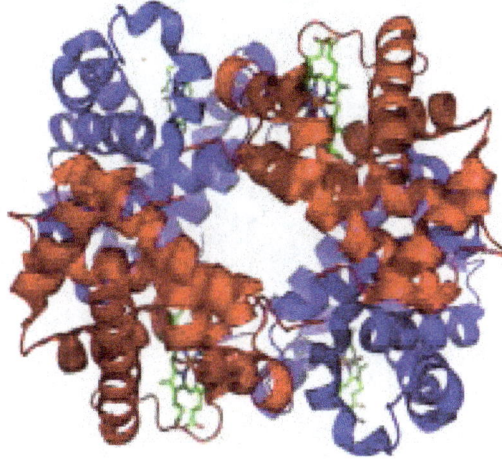

If the subunits are identical then it is a homomeric protein.

Heme

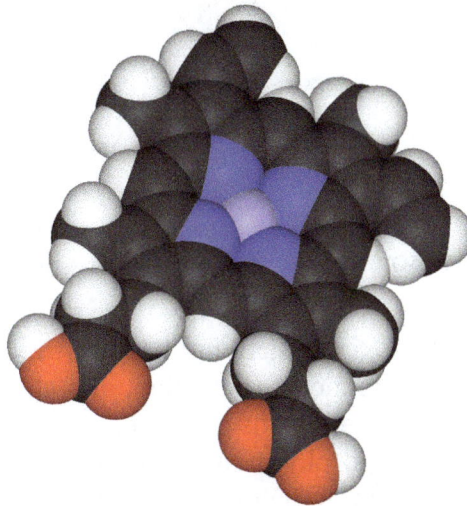

Space-filling model of heme B, with grey iron, blue nitrogen, black carbon, white hydrogen, red oxygen

Heme or haem is a cofactor consisting of an Fe^{2+} (ferrous) ion contained in the centre of a heterocyclic macrocycle organic compound called a porphyrin, made up of four pyrrolic groups joined together by methine bridges. Not all porphyrins contain iron, but a substantial fraction of porphyrin-containing metalloproteins have heme as their prosthetic group; these are known as hemoproteins. Hemes are most commonly recognized as components of hemoglobin, the red pigment in blood, but are also found in a number of other biologically important hemoproteins such as myoglobin, cytochrome, catalase, heme peroxidase, and endothelial nitric oxide synthase.

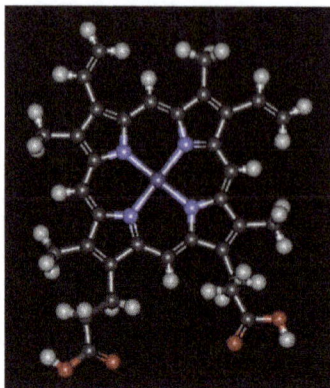

Ball and stick model of heme B

Function

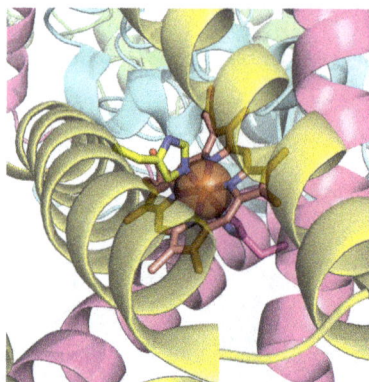

The heme group of succinate dehydrogenase bound to histidine, an electron carrier in the mitochondrial electron transfer chain. The large semi-transparent sphere indicates the location of the iron ion. From PDB: 1YQ3

Hemoproteins have diverse biological functions including the transportation of diatomic gases, chemical catalysis, diatomic gas detection, and electron transfer. The heme iron serves as a source or sink of electrons during electron transfer or redox chemistry. In peroxidase reactions, the porphyrin molecule also serves as an electron source. In the transportation or detection of diatomic gases, the gas binds to the heme iron. During the detection of diatomic gases, the binding of the gas ligand to the heme iron induces conformational changes in the surrounding protein. In general, diatomic gases only bind to the reduced heme, as ferrous $Fe(II)$ while most peroxidases cycle between $Fe(III)$ and $Fe(IV)$ and hemeproteins involved in mitochondrial redox, oxidation-reduction, cycle between $Fe(II)$ and $Fe(III)$.

It has been speculated that the original evolutionary function of hemoproteins was electron transfer in primitive sulfur-based photosynthesis pathways in ancestral cyanobacteria-like organisms before the appearance of molecular oxygen.

Hemoproteins achieve their remarkable functional diversity by modifying the environment of the heme macrocycle within the protein matrix. For example, the ability of hemoglobin to effectively deliver oxygen to tissues is due to specific amino acid residues located near the heme molecule. Hemoglobin reversibly binds to oxygen in the lungs when the pH is high, and the carbon dioxide concentration is low. When the situation is reversed (low pH and high carbon dioxide concentrations), hemoglobin will release oxygen into the tissues. This phenomenon, which states that hemoglobin's oxygen binding affinity is inversely proportional to both acidity and concentration of carbon dioxide, is known as the Bohr effect. The molecular mechanism behind this effect is the steric organization of the globin chain; a histidine residue, located adjacent to the heme group, becomes positively charged under acidic conditions (which are caused by dissolved CO_2 in working muscles, etc.) releasing oxygen from the heme group.

Types

Major Hemes

There are several biologically important kinds of heme:

		Heme A	Heme B	Heme C	Heme O
PubChem number		7888115	444098	444125	6323367
Chemical formula		$C_{49}H_{56}O_6N_4$-Fe	$C_{34}H_{32}O_4N_4$-Fe	$C_{34}H_{36}O_4N_4S_2$-Fe	$C_{49}H_{58}O_5N_4$-Fe
Functional group at C_3		$-CH(OH)$ CH_2Far	$-CH=CH_2$	$-CH(cyste-in-S-yl)CH_3$	$-CH(OH)$ CH_2Far
Functional group at C_8		$-CH=CH_2$	$-CH=CH_2$	$-CH(cyste-in-S-yl)CH_3$	$-CH=CH_2$
Functional group at C_{18}		$-CH=O$	$-CH_3$	$-CH_3$	$-CH_3$

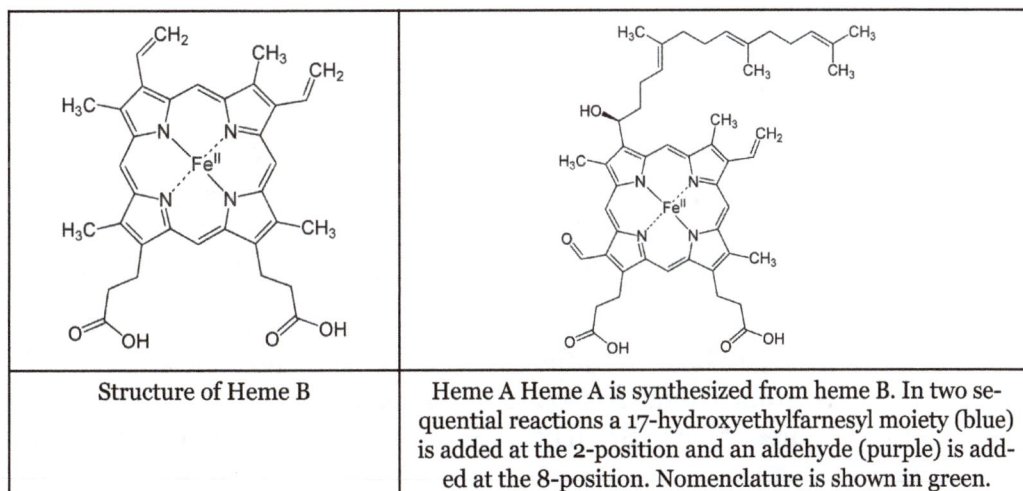

Structure of Heme B	Heme A Heme A is synthesized from heme B. In two sequential reactions a 17-hydroxyethylfarnesyl moiety (blue) is added at the 2-position and an aldehyde (purple) is added at the 8-position. Nomenclature is shown in green.

The most common type is *heme B*; other important types include *heme A* and *heme C*. Isolated hemes are commonly designated by capital letters while hemes bound to proteins are designated by lower case letters. Cytochrome a refers to the heme A in specific combination with membrane protein forming a portion of cytochrome c oxidase.

Other Hemes

The following carbon numbering system of porphyrins is an older numbering used by biochemists and not the 1–24 numbering system recommended by IUPAC which is shown in the table above.

- Heme l is the derivative of heme B which is covalently attached to the protein of lactoperoxidase, eosinophil peroxidase, and thyroid peroxidase. The addition of peroxide with the glutamyl-375 and aspartyl-225 of lactoperoxidase forms ester bonds between these amino acid residues and the heme 1- and 5-methyl groups, respectively. Similar ester bonds with these two methyl groups are thought to form in eosinophil and thyroid peroxidases. Heme l is one important characteristic of animal peroxidases; plant peroxidases incorporate heme B. Lactoperoxidase and eosinophil peroxidase are protective enzymes responsible for the destruction of invading bacteria and virus. Thyroid peroxidase is the enzyme catalyzing the biosynthesis of the important thyroid hormones. Because lactoperoxidase destroys invading organisms in the lungs and excrement, it is thought to be an important protective enzyme.

- Heme m is the derivative of heme B covalently bound at the active site of myeloperoxidase. Heme m contains the two ester bonds at the heme 1- and 5-methyls as in heme l found in other mammalian peroxidases. In addition, a unique sulfonium ion linkage between the sulfur of a methionyl amino-acid residue and the heme 2-vinyl group is formed, giving this enzyme the unique capability of easily oxidizing chloride and bromide ions. Myeloperoxidase is present in mammalian neutrophils and is responsible for the destruction of invading bacteria and viruse. It also synthesizes hypobromite by "mistake" which is a known mutagenic compound.

- Heme D is another derivative of heme B, but in which the propionic acid side chain at the carbon of position 6, which is also hydroxylated, forms a γ-spirolactone. Ring III is also hydroxylated at position 5, in a conformation trans to the new lactone group. Heme D is the site for oxygen reduction to water of many types of bacteria at low oxygen tension.

- Heme S is related to heme B by the having a formyl group at position 2 in place of the 2-vinyl group. Heme S is found in the hemoglobin of marine worms. The correct structures of heme B and heme S were first elucidated by German chemist Hans Fischer.

The names of cytochromes typically (but not always) reflect the kinds of hemes they contain: cytochrome a contains heme A, cytochrome c contains heme C, etc. This convention may have been first introduced with the publication of the structure of heme A.

Synthesis

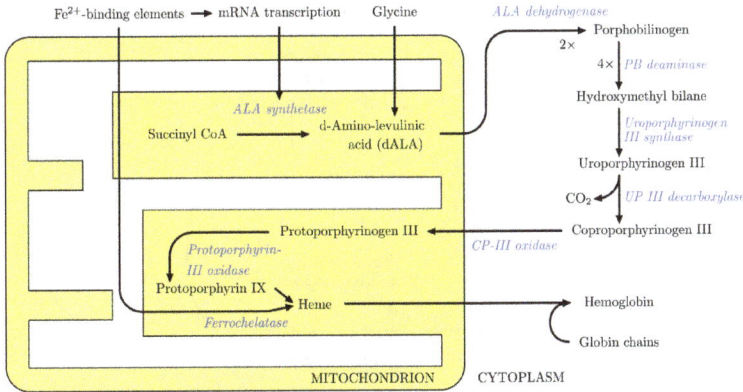

Heme synthesis in the cytoplasm and mitochondrion

The enzymatic process that produces heme is properly called porphyrin synthesis, as all the intermediates are tetrapyrroles that are chemically classified as porphyrins. The process is highly conserved across biology. In humans, this pathway serves almost exclusively to form heme. In other species, it also produces similar substances such as cobalamin (vitamin B_{12}).

The pathway is initiated by the synthesis of D-aminolevulinic acid (dALA or δALA) from the amino acid glycine and succinyl-CoA from the citric acid cycle (Krebs cycle). The rate-limiting enzyme responsible for this reaction, *ALA synthase*, is negatively regulated by glucose and heme concentration. Mechanism of inhibition of ALAs by heme or hemin is by decreasing stability of mRNA synthesis and by decreasing the intake of mRNA in the mitochondria. This mechanism is of therapeutic importance: infusion of *heme arginate* or *hematin* and glucose can abort attacks of acute intermittent porphyria in patients with an inborn error of metabolism of this process, by reducing transcription of ALA synthase.

The organs mainly involved in heme synthesis are the liver (in which the rate of synthesis is highly variable, depending on the systemic heme pool) and the bone marrow (in which rate of synthesis of Heme is relatively constant and depends on the production of globin chain), although every cell requires heme to function properly. However, due to its toxic properties, proteins such as Hemopexin (Hx) are required to help maintain physiological stores of iron in order for them to be used in synthesis. Heme is seen as an intermediate molecule in catabolism of hemoglobin in the process of bilirubin metabolism. Defects in various enzymes in synthesis of heme can lead to group of disorder called porphyrias, these include acute intermittent porphyria, congenital erythropoetic porphyria, porphyria cutanea tarda, hereditary coproporphyria, variegate porphyria, erythropoietic protoporphyria.

Degradation

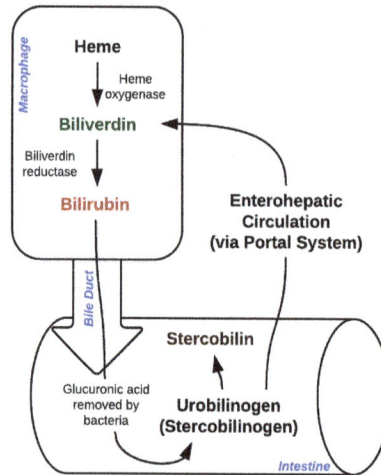

Heme breakdown

Degradation begins inside macrophages of the spleen, which remove old and damaged (senescent) erythrocytes from the circulation. In the first step, heme is converted to biliverdin by the enzyme heme oxygenase (HMOX). NADPH is used as the reducing agent, molecular oxygen enters the reaction, carbon monoxide (CO) is produced and the iron is released from the molecule as the ferrous ion (Fe^{2+}). CO acts as a cellular messenger and functions in vasodilation.

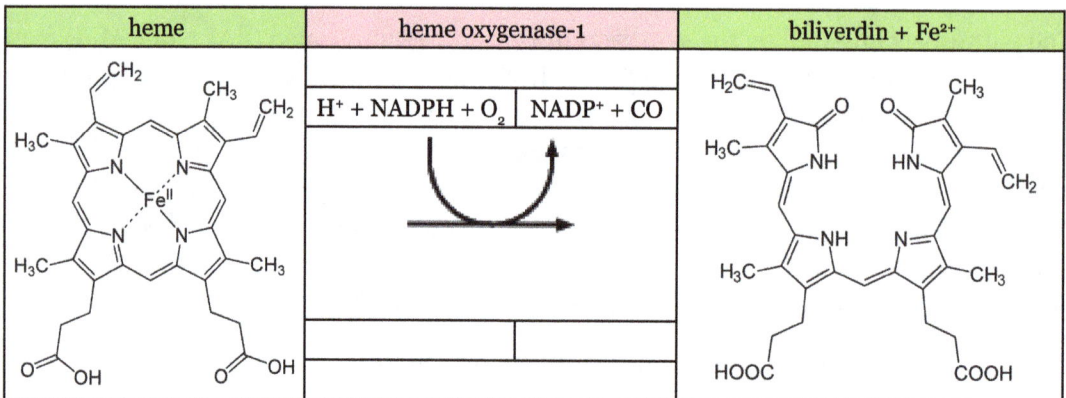

heme	heme oxygenase-1		biliverdin + Fe^{2+}
	$H^+ + NADPH + O_2$	$NADP^+ + CO$	

In addition, heme degradation appears to be an evolutionarily-conserved response to oxidative stress. Briefly, when cells are exposed to free radicals, there is a rapid induction of the expression of the stress-responsive heme oxygenase-1 (HMOX1) isoenzyme that catabolizes heme. The reason why cells must increase exponentially their capability to degrade heme in response to oxidative stress remains unclear but this appears to be part of a cytoprotective response that avoids the deleterious effects of free heme. When large amounts of free heme accumulates, the heme detoxification/degradation systems get overwhelmed, enabling heme to exert its damaging effects.

In the second reaction, biliverdin is converted to bilirubin by biliverdin reductase (BVR):

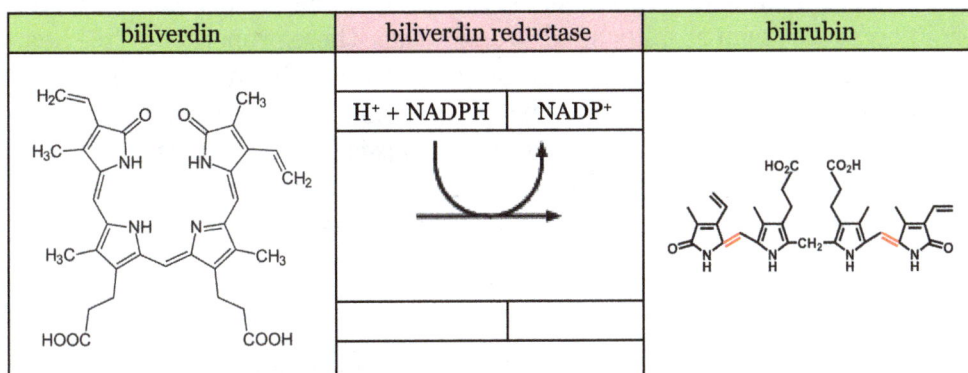

biliverdin	biliverdin reductase		bilirubin
	$H^+ + NADPH$	$NADP^+$	

Bilirubin is transported into the liver by facilitated diffusion bound to a protein (serum albumin), where it is conjugated with glucuronic acid to become more water-soluble. The reaction is catalyzed by the enzyme UDP-glucuronosyltransferase.

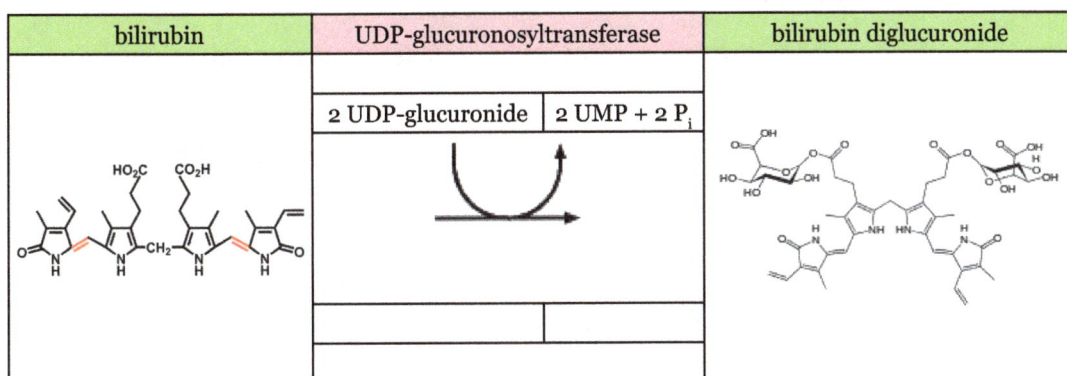

bilirubin	UDP-glucuronosyltransferase		bilirubin diglucuronide
	2 UDP-glucuronide	2 UMP + 2 P_i	

This form of bilirubin is excreted from the liver in bile. Excretion of bilirubin from liver to biliary canaliculi is an active, energy dependent and rate limiting process. The intestinal bacteria deconjugate bilirubin diglucuronide and convert bilirubin to uro-bilinogens. Some urobilinogen is absorbed by intestinal cells and transported into the kidneys and excreted with urine (urobilin, which is the product of oxidation of uro-bilinogen, is responsible for the yellow colour of urine). The remainder travels down the digestive tract and is converted to stercobilinogen. This is oxidized to stercobilin, which is excreted and is responsible for the color of feces.

In Health and Disease

Under homeostasis, the reactivity of heme is controlled by its insertion into the "heme pockets" of hemoproteins. Under oxidative stress however, some hemoproteins, e.g. he-moglobin, can release their heme prosthetic groups. The non-protein-bound (free) heme produced in this manner becomes highly cytotoxic, most probably due to the iron atom contained within its protoporphyrin IX ring, which can act as a Fenton's reagent to cata-lyze in an unfettered manner the production of free radicals. It catalyzes the oxidation and aggregation of protein, the formation of cytotoxic lipid peroxide via lipid peroxidation and damages DNA through oxidative stress. Due to its lipophilic properties, it impairs lipid bilayers in organelles such as mitochondria and nuclei. These properties of free heme can sensitize a variety of cell types to undergo programmed cell death in response to pro-inflammatory agonists, a deleterious effect that plays an important role in the pathogenesis of certain inflammatory diseases such as malaria and sepsis. There is an association between high intake of heme iron sourced from meat and increased risk of colon cancer. The heme content of red meat is 10-fold higher than that of white meat such as chicken.

Genes

The following genes are part of the chemical pathway for making heme:

- ALAD: aminolevulinic acid, δ-, dehydratase (deficiency causes ala-dehydratase deficiency porphyria)

- ALAS1: aminolevulinate, δ-, synthase 1

- ALAS2: aminolevulinate, δ-, synthase 2 (deficiency causes sideroblastic/hypochromic anemia)

- CPOX: coproporphyrinogen oxidase (deficiency causes hereditary coproporphyria)

- FECH: ferrochelatase (protoporphyria)

- HMBS: hydroxymethylbilane synthase (deficiency causes acute intermittent porphyria)

- PPOX: protoporphyrinogen oxidase (deficiency causes variegate porphyria)

- UROD: uroporphyrinogen decarboxylase (deficiency causes porphyria cutanea tarda)

- UROS: uroporphyrinogen III synthase (deficiency causes congenital erythropoietic porphyria)

Collagen, a fibrous protein   Myoglobin, a globular protein   Bacteriorhodopsin

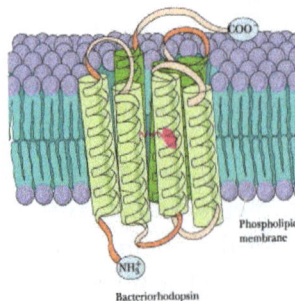

Fibrous Globular Membrane

Fibrous Proteins

Fibrous proteins are formed from long polypeptide chains that are arranged parallel or nearly parallel to one another. Fibrous polypeptide chains form long strands or sheets and because of many hydrophobic amino acid residues, they are water insoluble but strong and flexible. These long fibers or sheets result in physically tough materials that form the basis of structural proteins in the body. Examples of components that are comprised of such proteins are muscle, hair, nails etc.

- Fibrous proteins contain polypeptide chains organized parallel along a single axis, producing long fibers or large sheets;

- They are mechanically strong, play structural roles in nature;

- Difficult to dissolve in water.

Keratins and Collagen are Examples of Fibrous Proteins

α-keratins are found in hair, fingernails, claws, horns and beaks;

- Sequence consists of long alpha helical rod segments capped with non-helical N- and C-termini.

β-keratins are found in silk and consist of gly-ala repeat sequences;

- Ala is small and can be packed within the sheets.

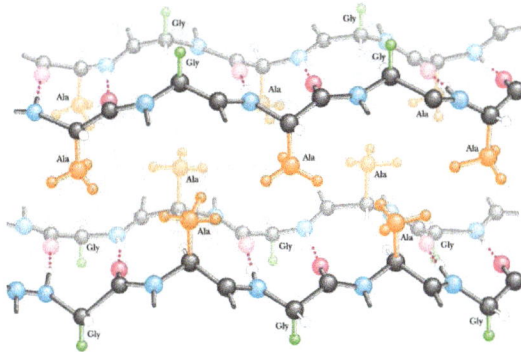

Globular Proteins

The polypeptide chain in this case folds into a compact structure close to a spherical shape. Most of these proteins are water soluble and because of this feature are mobile in the cell. They have diverse functions and act as enzymes and several regulatory proteins. Since globular proteins are compact mobile proteins they are also the major proteins that can function as transport proteins.

Globular proteins are classified according to the type and arrangement of secondary structure.

- Antiparallel alpha helix proteins

- Parallel or mixed beta sheet proteins

- Antiparallel beta sheet proteins

Membrane Proteins

These proteins are embedded in the lipid bilayer of the cell membrane and act as ion channels for the transport of molecules and ions in and out of the cell. Since the proteins interact with the non polar lipid bilayers, they have hydrophobic amino acid residues on the surface. These proteins do not have stable structures in aqueous solution.

Have you ever wondered what would happen if fibrous proteins were water soluble?

Residue	Globular protein	Membrane protein
Non-polar V L I M F Y W	In interior Hydrophobic core	Surface – lipid anchor
Polar charged R K D E H	Surface Catalytic sites	Hydrophilic core
Polar neutral S T N Q Y W	H bond network	Inside surface – part of channel

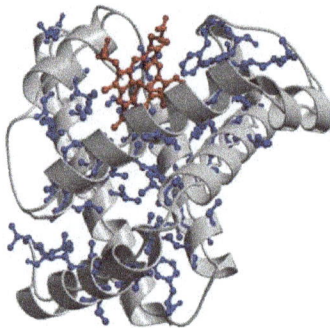

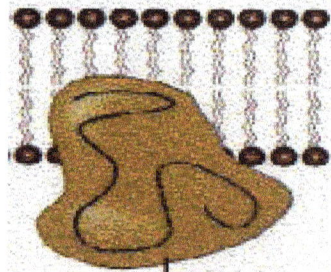

Some Protein Functions are Listed Below

Function	Example
Structure	Collagen in skin; keratin in hair, nails, horns
Movement	Actin and myosin in muscle
Defense	Antibodies in bloodstream
Storage	Albumin in egg white
Signaling	Growth hormone in bloodstream
Catalyzing reactions	Enzymes (Ex.: amylase digests carbohydrates; ATP synthase makes ATP)

Transport (hemoglobin) Transmembrane (Na+/K+ ATPases) Hormones (insulin)

Physico-Chemical Properties of Proteins: reflects amino acid composition and levels of organization

- AMPHOTERIC NATURE - ion transportation

- BUFFERING ABILITY - involves the imidazole group of HIS (pK = 6.1)

- SOLUBILITY - characteristic for each protein

  - at the isoelectric point (pI), we have zero charge

  - at pH on acid side of pI, then the net charge is (+)

  - at pH on basic side of pI, then the net charge is (-)

- SHAPE - very important

  - eg. enzyme recognition of a particular substrate.

References

- Paoli, M. (2002). "Structure-function relationships in heme-proteins.". DNA Cell Biol. 21 (4): 271–280. PMID 12042067. doi:10.1089/104454902753759690

- Alderton, W.K. (2001). "Nitric oxide synthases: structure, function and inhibition.". Biochem. J. 357 (3): 593–615. PMC 1221991. PMID 11463332. doi:10.1042/bj3570593

- Rae, T.; Goff, H. (1998). "The heme prosthetic group of lactoperoxidase. Structural characteristics of heme l and heme l-peptides". The Journal of Biological Chemistry. 273 (43): 27968–27977. PMID 9774411. doi:10.1074/jbc.273.43.27968

- Milani, M. (2005). "Structural bases for heme binding and diatomic ligand recognition in truncated hemoglobins.". J Inorg Biochem. 99 (1): 97–109. PMID 15598494. doi:10.1016/j.jinorgbio.2004.10.035

- Lehninger's Principles of Biochemistry (5th ed.). New York: W. H. Freeman and Company. 2008. p. 876. ISBN 978-0-7167-7108-1

- Poulos, T. (2014). "Heme Enzyme Structure and Function.". Chem. Rev. 114 (7): 3919–3962. PMC 3981943. PMID 24400737. doi:10.1021/cr400415k

- Murshudov, G.; Grebenko, A.; Barynin, V.; Dauter, Z.; Wilson, K.; Vainshtein, B.; Melik-Adamyan, W.; Bravo, J.; Ferrán, J.; Ferrer, J. C.; Switala, J.; Loewen, P. C.; Fita, I. (1996). "Structure of the heme d of Penicillium vitale and Escherichia coli catalases". The Journal of Biological Chemistry. 271 (15): 8863–8868. PMID 8621527. doi:10.1074/jbc.271.15.8863

- Thom, C. S. (2013). "Hemoglobin Variants: Biochemical Properties and Clinical Correlates.". Cold Spring Harb Perspect Med. 3 (3): a011858. PMC 3579210. PMID 23388674. doi:10.1101/cshperspect.a011858

- Ackers, G. K.; Holt, J. M. (2006). "Asymmetric cooperativity in a symmetric tetramer: human hemoglobin.". J Biol Chem. 281 (17): 11441–3. PMID 16423822. doi:10.1074/jbc.r500019200

- Plewinska, Magdalena; Thunell, Stig; Holmberg, Lars; Wetmur, James; Desnick, Robert (1991). "delta-Aminolevulinate dehydratase deficient porphyria: identification of the molecular lesions in a severely affected homozygote". American Journal of Human Genetics. 49 (1): 167–174. PMC 1683193. PMID 2063868

- Hegg, Eric L. (2004). "Heme A Synthase Does Not Incorporate Molecular Oxygen into the Formyl Group of Heme A". Biochemistry. 43 (27): 8616–8624. PMID 15236569. doi:10.1021/bi049056m

- Yoshikawa, S. (2012). "Structural studies on bovine heart cytochrome c oxidase". Biochem Biophys Acta. 1817 (4): 579–589. PMID 22236806. doi:10.1016/j.bbabio.2011.12.012

Protein Folding and Purification

Proteins change their color or structure when some external pressure is applied on them. They must be folded in the correct shape for them to function properly. The common example cited in the denaturing of proteins is that of egg whites. The topics discussed in the chapter are of great importance to broaden the existing knowledge on protein chemistry.

Protein Folding

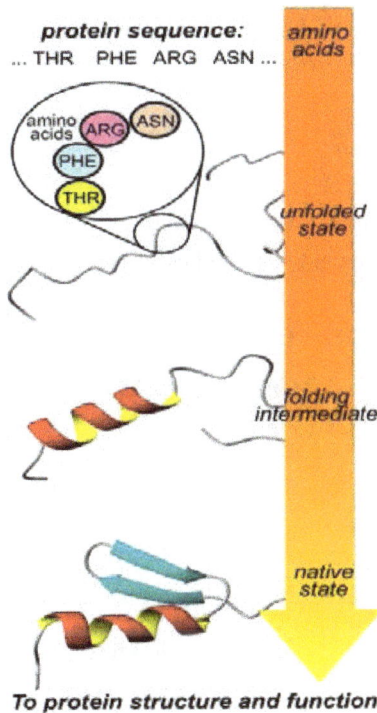

protein sequence:
... THR PHE ARG ASN ...

amino acids

amino acids (ARG) (ASN) (PHE) (THR)

unfolded state

folding intermediate

native state

To protein structure and function

Proteins are synthesized in the ribosome as a linear sequence of amino- acids. The newly synthesized polypeptide folds into its characteristic and functional three-dimensional structure via a physical process known as protein folding. Several interactions among the amino acids lead to the formation of a folded three dimensional structure which is known as the native state of the protein. We know that the three dimensional structure is determined by its amino acid sequence as discussed earlier.

The failure of a protein to be able to fold into its native structure produces a misfolded protein which has oftentimes been correlated with disease because the protein cannot perform its designated function.

It is believed that hydrophobic collapse is a key driving force for protein folding resulting in the formation of a hydrophobic core with the polar surface interacting with the solvent.

Models for Protein Folding: Framework Model

This model proposes native-like microdomain formation (α -helices, β -hairpins, etc.) as a result of the folding reaction. These small secondary structural units formed during the early stages of protein folding are together able to construct a final stable tertiary structure with native-like contacts.

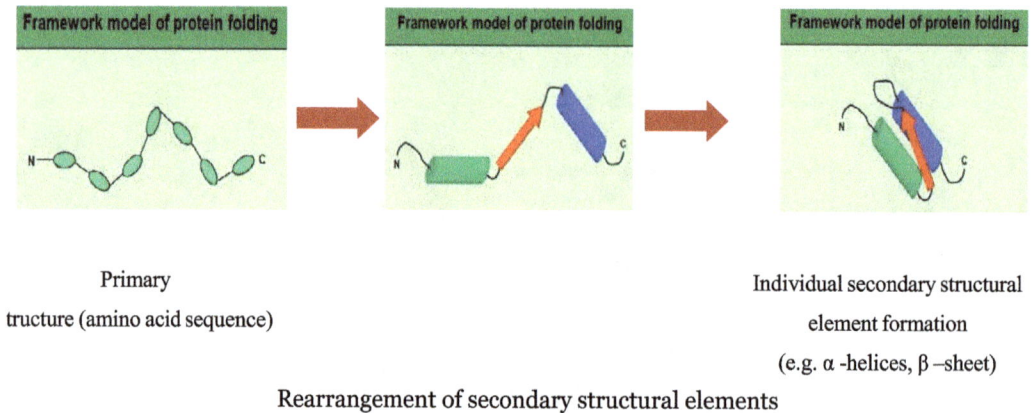

Primary

tructure (amino acid sequence)

Individual secondary structural element formation

(e.g. α -helices, β –sheet)

Rearrangement of secondary structural elements

Nucleation and Nucleation-condensation Model:

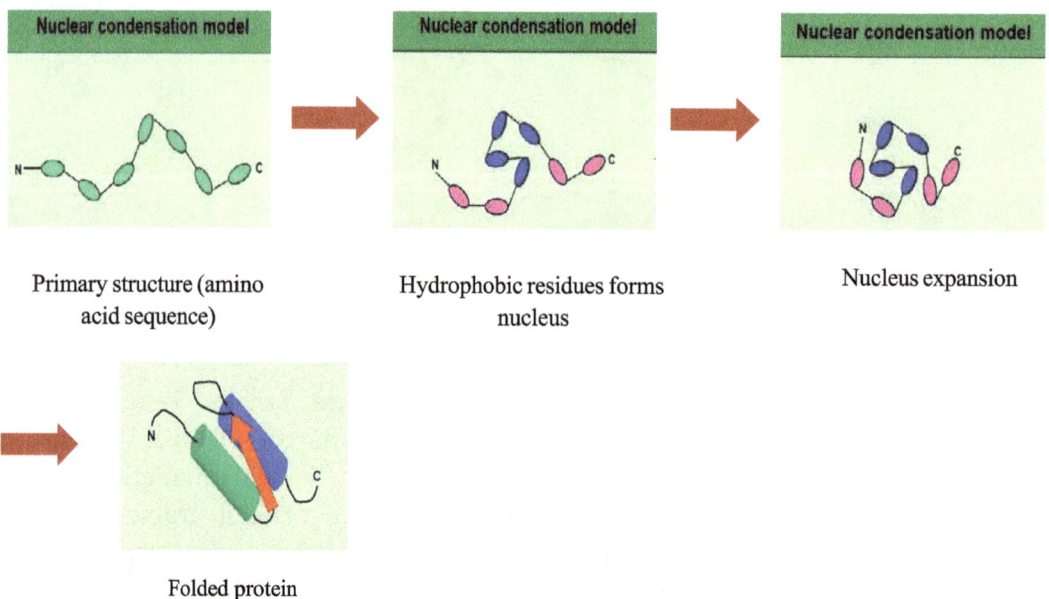

Primary structure (amino acid sequence)

Hydrophobic residues forms nucleus

Nucleus expansion

Folded protein

According to this model formation of a confined nucleus of secondary structure takes place via some specific residues of the polypeptide chain in the rate-limiting step of folding. The complete native structure shapes up around this nucleus. A similar type of idea is in the nucleation– condensation model which states that the nucleus of local secondary structure requires interactions with non-local residues for its stabilization as a result of its poor stability. Thus the initial formation of a diffuse folding nucleus is followed by the formation of secondary structure along with tertiary interactions that take place in a concerted step.

Molten Globule Model

This model hypothesizes that folding proceeds via an initial clustering among side chains of hydrophobic residue which prefer to be aloof from an aqueous environment. The process of clustering occurs rapidly due to non specific interaction among the hydrophobic residues that lead to the formation of a relatively compact arrangement (molten globule state). Hydrophobic residues of the proteins gather inside the collapsed forms within the core. The collapsed state favors the formation of secondary structure and encourages tertiary interaction among the residues.

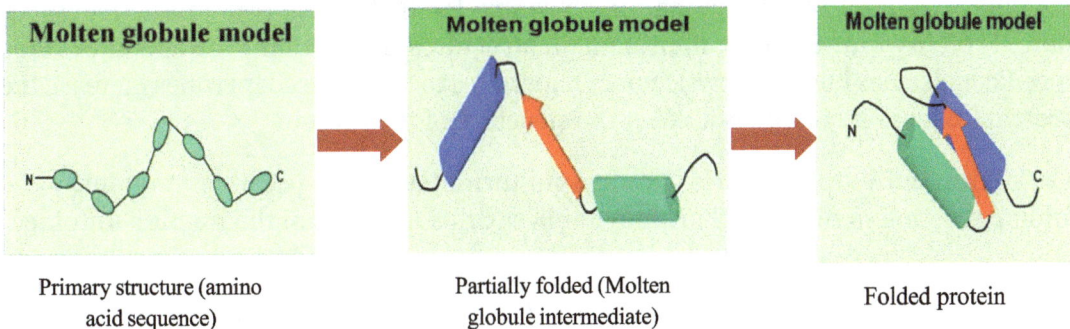

| Primary structure (amino acid sequence) | Partially folded (Molten globule intermediate) | Folded protein |

Energy Landscape Theory (Folding Funnel)

Fundamental constraints of protein structures, φ and ψ angles dictate possible polypeptide backbone conformations. Both local and nonlocal interactions play a role in protein folding. Apart from these, hydrophobic interactions and hydrogen bonding play an important role.

The folding funnel model postulates that protein molecules navigate via a funnel-shaped energy landscape during the folding process. The folding funnel represents a plot of the enthalpy versus configurational entropy. A funnel-shaped energy landscape suggests that the structure which resembles the native structure will possess the lowest free energy. The top of the funnel is broad and represents the denatured state with high conformational entropy.

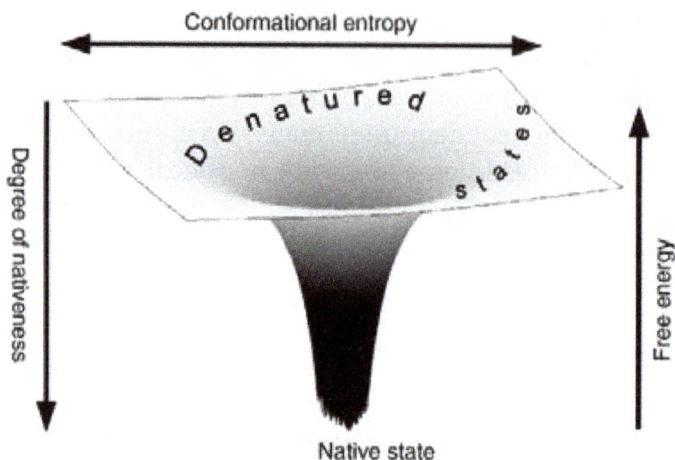

It becomes narrow towards the bottom as the structure acquires a native like structure with low conformational entropy.

Thermodynamics and Equilibrium of Protein Folding

The folded state of protein is marginally stable (5–10 kcal/mol) than the denatured state. Proteins with complex multidomain structures may require the action of some specific molecules known as molecular chaperones to fold. The chaperones increase the protein folding rate but do not affect the structure of the protein.

The native and unfolded states are in equilibrium and it is possible to quantify the folding reaction in terms of thermodynamics with a measure of the fraction unfolded.

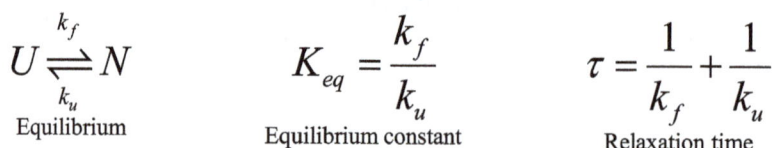

$$U \underset{k_u}{\overset{k_f}{\rightleftharpoons}} N$$

Equilibrium

$$K_{eq} = \frac{k_f}{k_u}$$

Equilibrium constant

$$\tau = \frac{1}{k_f} + \frac{1}{k_u}$$

Relaxation time

N=Native (fully folded); U=Unfolded (denatured)

Free energy of Folding

Gibbs free Energy of Folding is Defined as

$$\Delta G_{NU} = G_{N\_}\Delta G_U = -RT \ lnK_{eq}$$

ΔG : more negative free energy more stable system.

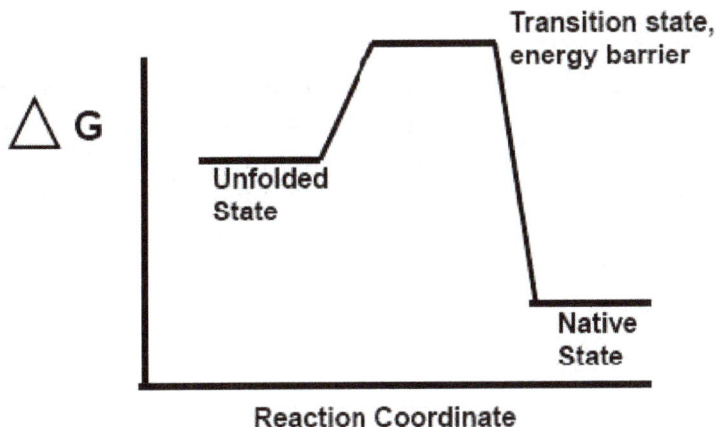

Reaction Coordinate

Entropy and Enthalpy of Folding

$$\Delta G_{NU} = \Delta H - T\Delta S$$

The value of enthalpy (ΔH) is negative for both states and small for the unfolded protein and relatively large for folded proteins due to the number of interactions between the amino acid residues in the folded state.

The entropy decreases as the protein attains its three-dimensional native structure, since it goes from an unfolded state to a compact structure. However the entropy change should also be considered in terms of the water molecules that surround the unfolded state and how their entropy changes when the protein folding process occurs. In general the entropy change is negative that makes the contribution of the $T\Delta S$ term to the overall ΔG positive.

Since the value of ΔH is negative, this compensates for the positive contribution from the ΔS term making the overall folding process of the protein enthalpy driven.

How do we Know a Protein has Folded Correctly?

Experimental Techniques for Studying Protein Folding

Mutation studies provide specific information regarding amino acid residues protein folding. In addition to this various experimental techniques provide fruitful information about protein folding and unfolding phenomena.

Protein Nuclear Magnetic Resonance Spectroscopy

NMR spectroscopy has been frequently used to study protein folding. The overall stability and residue-specific stability can be determined by monitoring hydrogen-deuterium exchange of backbone amide protons of proteins in their native state.

Circular Dichroism

The ability of protein to absorb circularly polarized light due to its chiral structural contents such as alpha helices and beta sheets makes circular dichroism an effective tool to study protein folding. The folding equilibrium of protein can be determined by measuring the change in absorption of protein as a function of denaturant concentration or change in temperature. The utility of this type of spectroscopy can be improved by combining it with fast-mixing devices, for instance stopped flow, to measure protein folding kinetics.

Dual Polarization Interferometry

Optical properties of molecular layers can be measured using a surface based technique known as dual polarization interferometry. By determining the overall size of protein monolayer and its density in real time at sub-Angstrom resolution one can characterize protein folding. Although it is limited for the processes slower than ~10 Hz. The incentive for folding can be a denaturant or temperature.

Vibrational Circular Dichroism of Proteins

The Fourier transform vibrational circular dichroism (VCD) instrument is an effective tool for determining protein conformations in solution even for very large protein molecules. VCD along with X-ray diffraction of protein crystals and FT-IR data for protein solutions in heavy water (D2O), provide explicit structural information that is not possible from CD measurements.

Time Resolved Techniques

With the improvement of time resolved techniques, the study of protein folding has gained importance. Time resolved studies involve methods which rapidly prompt the folding of an unfolded protein followed by monitoring of consequential dynamics. A few examples of these techniques are neutron scattering, ultrafast mixing of solutions, photochemical methods, and laser temperature jump spectroscopy.

Proteolysis

This technique involves probing the fraction of unfolded protein under various solution conditions.

Molecular Chaperones in Folding within the Cell

Many proteins after removal of denaturants are known to refold *in vitro*. This suggest that amino acid sequence carries all the necessary information to rebulild thre dimmensional protein structure. However reports have suggested that under cellular environment many protein requires assistence of biologically active molecule for their folding process. These biologically active molecules are commonly known as molecular chaperones. Some proteins are used to bind with unfolded protein to avoid misfolding (e.g. DnaJ). There exist another class of molecules which capture misfolded proteins and unfold them using chemical energy they are known as chaperonines (e.g. GroEL-GroES).

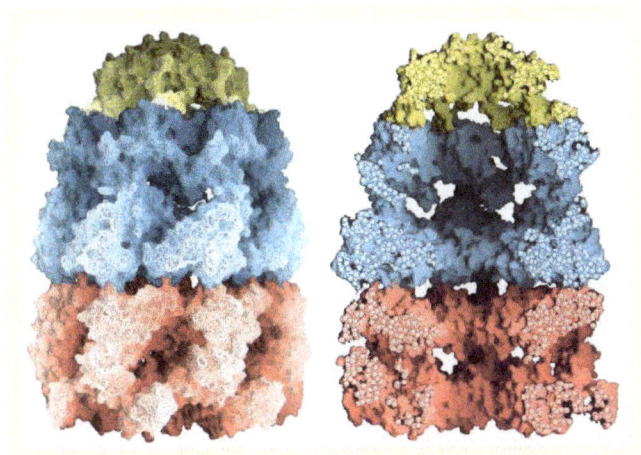

The GroEL-GroES Complex is a Two-Stroke Motor with Unfoldase Activity

Denaturation (Biochemistry)

Denaturation is a process in which proteins or nucleic acids lose the quaternary structure, tertiary structure and secondary structure which is present in their native state, by application of some external stress or compound such as a strong acid or base, a concentrated inorganic salt, an organic solvent (e.g., alcohol or chloroform), radiation or heat. If proteins in a living cell are denatured, this results in disruption of cell activity and possibly cell death. Protein denaturation is also a consequence of cell death. Denatured proteins can exhibit a wide range of characteristics, from conformational change and loss of solubility to aggregation due to the exposure of hydrophobic groups.

Protein folding is key to whether a globular protein or a membrane protein can do its job correctly. It must be folded into the right shape to function. But hydrogen bonds, which play a big part in folding, are rather weak, and it doesn't take much heat, acidity, varying salt concentrations, or other stress to break some and form others, denaturing the protein. This is one reason why tight homeostasis is physiologically necessary in many life forms.

This concept is unrelated to denatured alcohol, which is alcohol that has been mixed with additives to make it unsuitable for human consumption.

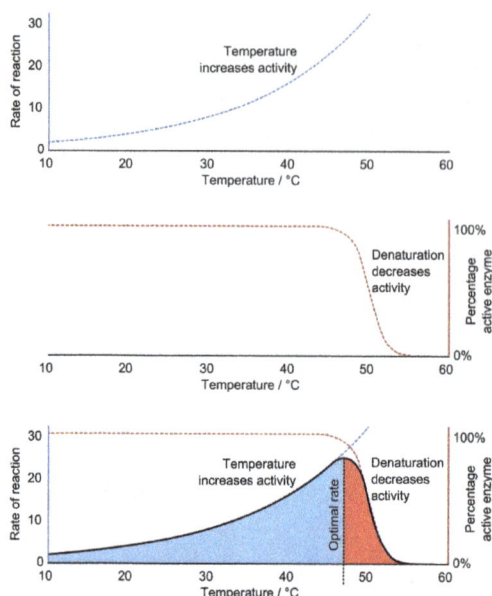

The effects of temperature on enzyme activity. Top - increasing temperature increases the rate of reaction (Q10 coefficient). Middle - the fraction of folded and functional enzyme decreases above its denaturation temperature. Bottom - consequently, an enzyme's optimal rate of reaction is at an intermediate temperature.

IUPAC Definition

Process of partial or total alteration of the native secondary, and/or tertiary, and/or quaternary structures of proteins or nucleic acids resulting in a loss of *bioactivity*.

Note 1: Modified from the definition.

Note 2: Denaturation can occur when proteins and nucleic acids are subjected to elevated temperature or to extremes of pH, or to nonphysiological concentrations of salt, organic solvents, urea, or other chemical agents.

Note 3: An *enzyme* loses its catalytic activity when it is denaturized.

Common Examples

When food is cooked, some of its proteins become denatured. This is why boiled eggs become hard and cooked meat becomes firm.

A classic example of denaturing in proteins comes from egg whites, which are typically largely egg albumins in water. Fresh from the eggs, egg whites are transparent and liquid. Cooking the thermally unstable whites turns them opaque, forming an interconnected solid mass. The same transformation can be effected with a denaturing chemical. Pouring egg whites into a beaker of acetone will also turn egg whites translucent and solid. The skin that forms on curdled milk is another common example of denatured protein. The cold appetizer known as ceviche is prepared by chemically "cooking" raw fish and shellfish in an acidic citrus marinade, without heat.

(Top) The protein *albumin* in the egg white undergoes denaturation and loss of solubility when the egg is cooked. (Bottom) Paperclips provide a visual analogy to help with the conceptualization of the denaturation process.

Protein Denaturation

Denatured proteins can exhibit a wide range of characteristics, from loss of solubility to protein aggregation.

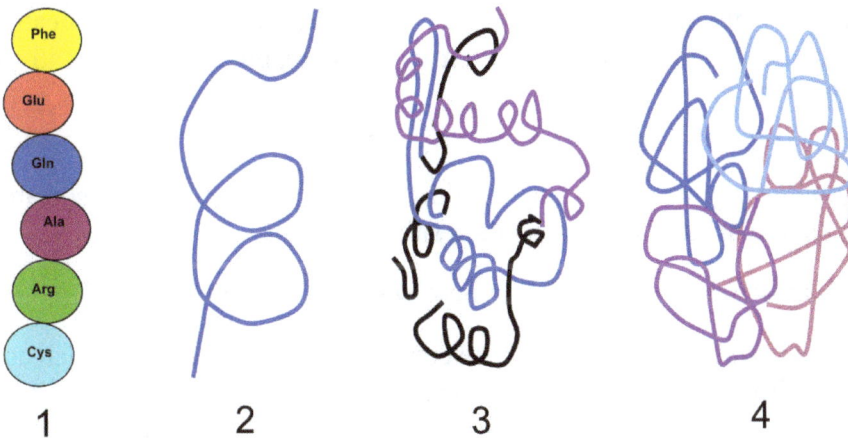

Functional proteins have four levels of structural organization:
1) Primary Structure : the linear structure of amino acids in the polypeptide chain
2) Secondary Structure : hydrogen bonds between peptide group chains in an alpha helix or beta sheet
3) Tertiary Structure : three-dimensional structure of alpha helixes and beta helixes folded
4) Quaternary Structure : three-dimensional structure of multiple polypeptides and how they fit together

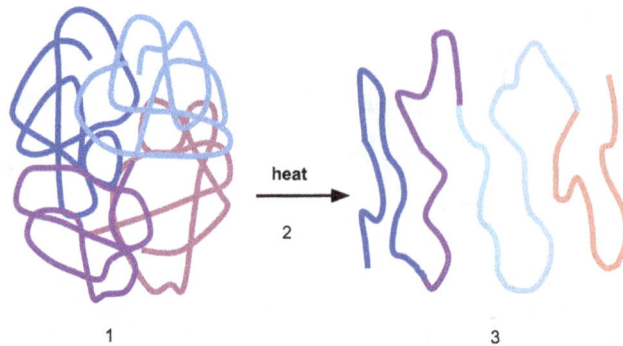

Process of Denaturation: 1) Functional protein showing a quaternary structure 2) when heat is applied it alters the intramolecular bonds of the protein 3) unfolding of the polypeptides (amino acids)

Background

Proteins are amino acid polymers. A protein is created by ribosomes that "read" RNA that is encoded by codons in the gene and assemble the requisite amino acid combination from the genetic instruction, in a process known as translation. The newly created protein strand then undergoes posttranslational modification, in which additional atoms or molecules are added, for example copper, zinc, or iron. Once this post-translational modification process has been completed, the protein begins to fold (sometimes spontaneously and sometimes with enzymatic assistance), curling up on itself so that hydrophobic elements of the protein are buried deep inside the structure and hydrophilic elements end up on the outside. The final shape of a protein determines how it interacts with its environment.

When a protein is denatured, secondary and tertiary structures are altered but the peptide bonds of the primary structure between the amino acids are left intact. Since all structural levels of the protein determine its function, the protein can no longer perform its function once it has been denatured. This is in contrast to intrinsically unstructured proteins, which are unfolded in their native state, but still functionally active.

How Denaturation Occurs at Levels of Protein Structure

- In quaternary structure denaturation, protein sub-units are dissociated and/or the spatial arrangement of protein subunits is disrupted.

- Tertiary structure denaturation involves the disruption of:

    o Covalent interactions between amino acid side-chains (such as disulfide bridges between cysteine groups).

    o Non-covalent dipole-dipole interactions between polar amino acid side-chains (and the surrounding solvent).

    o Van der Waals (induced dipole) interactions between nonpolar amino acid side-chains.

- In secondary structure denaturation, proteins lose all regular repeating patterns such as alpha-helices and beta-pleated sheets, and adopt a random coil configuration.

- Primary structure, such as the sequence of amino acids held together by covalent peptide bonds, is not disrupted by denaturation.

Loss of Function

Most biological substrates lose their biological function when denatured. For example, enzymes lose their activity, because the substrates can no longer bind to the active site, and because amino acid residues involved in stabilizing substrates' transition states are no longer positioned to be able to do so. The denaturing process and the associated loss of activity can be measured using techniques such as dual polarization interferometry, CD, QCM-D and MP-SPR.

Reversibility and Irreversibility

In very few cases, denaturation is reversible (the proteins can regain their native state when the denaturing influence is removed). This process can be called renaturation. This understanding has led to the notion that all the information needed for proteins to assume their native state was encoded in the primary structure of the protein, and hence in the DNA that codes for the protein, the so-called "Anfinsen's thermodynamic hypothesis".

Nucleic Acid Denaturation

Nucleic acids (including RNA and DNA) are nucleotide polymers synthesized by polymerase enzymes during either transcription or DNA replication. Following 5'-3' synthesis of the backbone, individual nitrogenous bases are capable of interacting with one another via hydrogen bonding, thus allowing for the formation of higher-order structures. Nucleic acid denaturation occurs when hydrogen bonding between nucleotides is disrupted, and results in the separation of previously annealed strands. For example, denaturation of DNA due to high temperatures results in the disruption of Watson and Crick base pairs and the separation of the double stranded helix into two single strands. Nucleic acid strands are capable of re-annealling when "normal" conditions are restored, but if restoration occurs too quickly, the nucleic acid strands may re-anneal imperfectly resulting in the improper pairing of bases.

Biologically-Induced Denaturation

The non-covalent interactions between antiparallel strands in DNA can be broken in order to "open" the double helix when biologically important mechanisms such as DNA replication, transcription, DNA repair or protein binding are set to occur. The area of partially separated DNA is known as the denaturation bubble, which can be more specifically defined as the opening of a DNA double helix through the coordinated separation of base pairs.

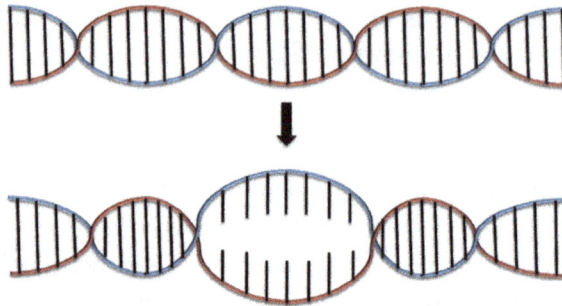

DNA denaturation occurs when hydrogen bonds between Watson and Crick base pairs are disturbed.

The first model that attempted to describe the thermodynamics of the denaturation bubble was introduced in 1966 and called the Poland-Scheraga Model. This model describes the denaturation of DNA strands as a function of temperature. As the temperature increases, the hydrogen bonds between the Watson and Crick base pairs are increasingly disturbed and "denatured loops" begin to form. However, the Poland-Scheraga Model is now considered elementary because it fails to account for the confounding implications of DNA sequence, chemical composition, stiffness and torsion.

Recent thermodynamic studies have inferred that the lifetime of a singular denaturation bubble ranges from 1 microsecond to 1 millisecond. This information is based on established timescales of DNA replication and transcription. Currently, biophysical and biochemical research studies are being performed to more fully elucidate the thermodynamic details of the denaturation bubble.

Denaturation Due to Chemical Agents

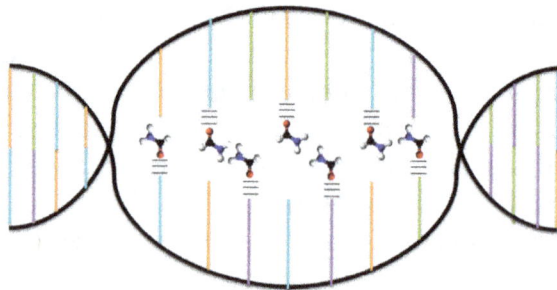

Formamide denatures DNA by disrupting the hydrogen bonds between Watson and Crick base pairs. Orange, blue, green, and purple lines represent adenine, thymine, guanine, and cytosine respectively. The three short black lines between the bases and the formamide molecules represent newly formed hydrogen bonds.

With polymerase chain reaction (PCR) being among the most popular contexts in which DNA denaturation is desired, heating is the most frequent method of denaturation. Other than denaturation by heat, nucleic acids can undergo the denaturation process through various chemical agents such as formamide, guanidine, sodium salicylate, dimethyl sulfoxide (DMSO), propylene glycol, and urea. These chemical denaturing agents lower the melting temperature (T_m) by competing for hydrogen bond donors and acceptors with pre-existing nitrogenous base pairs. Some agents are even able to induce denaturation at room temperature. For example, alkaline agents (e.g. NaOH) have been shown to denature DNA by changing pH and removing hydrogen-bond contributing protons. These denaturants have been employed to make Denaturing Gradient Gel Electrophoresis gel (DGGE), which promotes denaturation of nucleic acids in order to eliminate the influence of nucleic acid shape on their electrophoretic mobility.

Chemical Denaturation as an Alternative

The optical activity (absorption and scattering of light) and hydrodynamic properties (translational diffusion, sedimentation coefficients, and rotational correlation times) of formamide denatured nucleic acids are similar to those of heat-denatured nucleic acids. Therefore, depending on the desired effect, chemically denaturing DNA can provide a gentler procedure for denaturing nucleic acids than denaturation induced by heat. Studies comparing different denaturation methods such as heating, beads mill of different bead sizes, probe sonication, and chemical denaturation show that chemical denaturation can provide quicker denaturation compared to the other physical denaturation methods described. Particularly in cases where rapid renaturation is desired, chemical denaturation agents can provide an ideal alternative to heating. For example, DNA strands denatured with alkaline agents such as NaOH renature as soon as phosphate buffer is added.

Denaturation Due to Air

Small, electronegative molecules such as nitrogen and oxygen, which are the primary gases in air, significantly impact the ability of surrounding molecules to participate in hydrogen bonding. These molecules compete with surrounding hydrogen bond acceptors for hydrogen bond donors, therefore acting as "hydrogen bond breakers" and weakening interactions between surrounding molecules in the environment. Antiparellel strands in DNA double helices are non-covalently bound by hydrogen bonding between Watson and Crick base pairs; nitrogen and oxygen therefore maintain the potential to weaken the integrity of DNA when exposed to air. As a result, DNA strands exposed to air require less force to separate and exemplify lower melting temperatures.

Applications

Many laboratory techniques rely on the ability of nucleic acid strands to separate. By understanding the properties of nucleic acid denaturation, the following methods were created:

- PCR
- Southern blot
- Northern blot
- DNA Sequencing

Denaturants

Protein Denaturants

Acids

Acidic protein denaturants include:

- Acetic acid
- Trichloroacetic acid 12% in water
- Sulfosalicylic acid

Bases

Bases work similarly to acids in denaturation. They include:

- Sodium bicarbonate

Solvents

Most organic solvents are denaturing, including:

- Ethanol
- alcohol

Cross-linking Reagents

Cross-linking agents for proteins include:

- Formaldehyde
- Glutaraldehyde

Chaotropic Agents

Chaotropic agents include:

- Urea 6 – 8 mol/l
- Guanidinium chloride 6 mol/l
- Lithium perchlorate 4.5 mol/l

Disulfide Bond Reducers

Agents that break disulfide bonds by reduction include:

- 2-Mercaptoethanol
- Dithiothreitol
- TCEP (tris(2-carboxyethyl)phosphine)

Other

- Mechanical agitation
- Picric acid
- Radiation
- Temperature

Nucleic Acid Denaturants

Chemical

Acidic nucleic acid denaturants include:

- Acetic acid
- HCl

Basic nucleic acid denaturants include:

- NaOH

Other nucleic acid denaturants include:

- DMSO
- Formamide
- Guanidine
- sodium salicylate
- Propylene glycol
- Urea

Physical

- Thermal denaturation

- Beads mill

- Probe sonification

- Radiation

Considering the forces involved in the formation of the tertiary structure of the protein, we realize that several factors are able to affect the folding process. Disruption of the forces by external agents can in turn lead to a disruption of the protein structure.

The process in which any physical or chemical factor results in the loss of the native structure of the protein molecule with a corresponding loss of biological activity is termed Protein Denaturation.

The process of denaturation of proteins thus destroys the colloidal nature of the protein thus rendering them incapable of interactions with water. This can result in precipitation and the may be permanent or temporary.

A very common example of protein denaturation is that of poaching an egg. The albumin of the egg which is a protein denatures upon heating.

Thus the disruption of the noncovalent interactions will therefore lead to unfolding of the protein and can lead to reversible or irreversible denaturation.

Native state (N)   Denatured state (D)

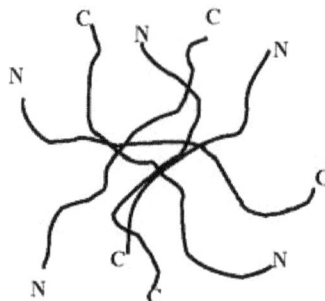

$$\Delta G_D^N = -RTIn[D]/[N]$$

The free energy on going from the native (N) state to the denatured (D) state is given by $\Delta G_N^D = \Delta H_N^D - T\Delta S_N^D$.

The overall free energy change ΔG_D^N depends on the combined effects of

-    the exposure of the non-polar groups to the solvent

-    the disruption in the ordered structure of the protein and

-    the interactions of the side chains with water

Factors that cause denaturation are

(i)   Temperature

(ii)   pH

(iii)   Electrolyte addition

(iv)   Denaturants such as urea and Guanidinium Hydrochloride (GuHCl)

(v)   Organic solvents

Each of the factors mentioned are capable of disrupting a specific force or forces that are responsible for holding the overall structure of the protein together.

We will see the major cause of action of the specific denaturing agents.

1) Temperature

In this case greater energy is supplied to the system in the form of heat energy which is capable of breaking bonds.

2) pH

Drastic changes in pH are able to affect the charge on the protein and change its ability to interact with water.

3) Electrolyte addition

The addition of salt is similar to the process of salting out which we will learn later. In this case the salt concentration interferes with the colloidal state of the protein in most cases.

4) Denaturants such as urea and Guanidinium Hydrochloride (GuHCl)

These commonly known denaturants are able to disrupt the hydrogen bonding network that is important in keeping the three-dimensional structure ofthe protein intact.

5) Organic solvents

The organic solvents are those that most often interfere with the dielectric constant of the medium.

There are many ways that may be used to investigate the denaturation of a protein.

The most common methods used are those that show a distinct change in some property of the protein that can be monitored. For example, UV measurements can measure the extent of unfolding with a denaturant provided there are aromatic amino acid residues that will be able to absorb the UV light. This is also the case with fluorescence. Changes in secondary structure of the protein can be monitored through CD experiments. This will measure the disruption in the secondary structure of the protein upon denaturation. Viscosity measurements can also give an insight into how the protein molecule behaves on unfolding as the compact protein will denature and open up forming a network that in practically all cases causes an increase in the relative viscosity of the protein in presence of increasing amounts of the denaturant.

DETERMINATION OF $\triangle G^{\circ}_{Dwater}$ FOR PROTEIN DENATURATION

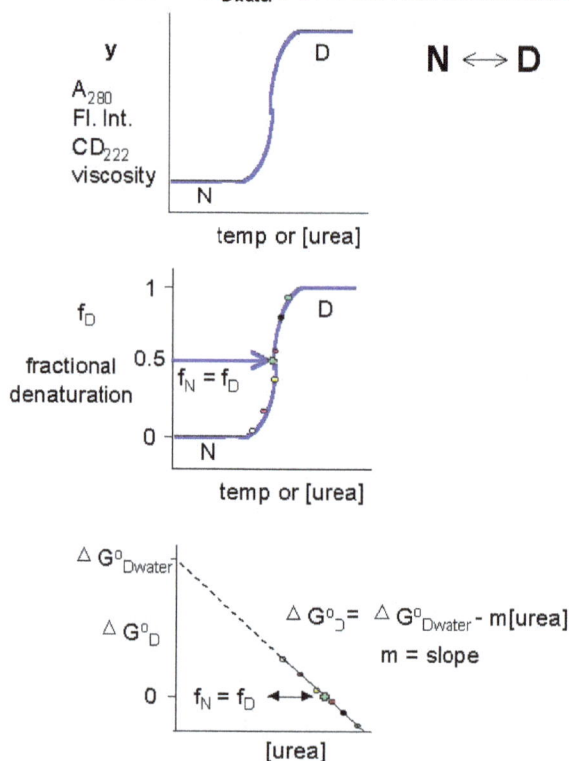

In all these methods the measurable dependent variable (y) is monitored as a function of the causative agent which in this case is a measure of the denaturant (temperature changes or changes in urea concentration etc.).

The denaturation curves normally exhibit a sigmoidal curve that is indicative of the co-operative transition from the native to the denatured state. It is also possible to obtain a measure of the fractional denaturation.

The equation used for this purpose is:

$$\Delta G = -RT\ln K = -RT\ln\left[\left(yN - y\right)/\left(y - yD\right)\right]$$

where K is the equilibrium constant, y is the observed value of the parameter used to follow unfolding, yN and yD are the values of y characteristic of the native and denatured conformations respectively (these are functions of urea concentration: y = a + b[urea]).

Misfolding and Neurodegenerative Disease

The misfolding of protein leads to formation of aggregated insoluble extracellular deposits which are known to cause several lethal neurodisorders such as Alzheimer's, Parkinson's, and Huntington's. These aggregated proteins have characteristic crossed β-sheet structure and are commonly known as amyloid fibrils. However it is still unclear that whether the fibrillar aggregates or the intermediate formed in the process of amyloid formation are the causative agents. The reduction in the number of functional protein molecules due to misfolding and excessive degradation, results in loss of the normal cellular function which is responsible for diseases such as cystic fibrosis, lysosomal storage diseases and early-onset emphysema.

The protein replacement therapy has been used in the past to correct the disorder but use of pharmaceutical chaperones to refold the mutated protein is a fast growing method to render them functional.

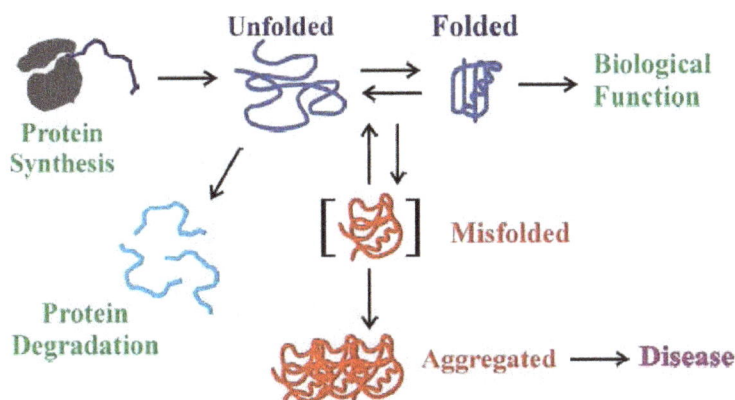

We understand therefore that fundamental constraints of protein structures, φ and ψ angles dictate possible polypeptide backbone conformations. Both local and nonlocal interactions play a role in protein folding. Apart from these hydrophobic interactions and hydrogen bonding play important roles

Protein Aggregation

A functional protein adopts a three dimensional structure due to folding of a linear sequence of amino acids which is considered to be one of the most essential biological processes. The number of naturally occurring amino acids is 20 which have broadened the diversity of and range of possible amino acids sequences. Folding of proteins into a rigid unique three dimensional structure is responsible for various functions exhibited by protein *in vivo*. Thus the relationship between structure and functions of proteins and their link to various cellular processes has evolved as a rising issue in recent times which spans both the unstructured proteins (functional) and unstructured regions in functional proteins (Dyson and Wright, 2005). There is an intimate relationship between protein aggregation, folding and stability of proteins.

Protein aggregation can be classified in the following categories (Fink, 1998):

- *In vivo* aggregation

- *In vitro* aggregation

- Formation of ordered structures (eg. Amyloid aggregates both *in vivo* and *in vitro*)

- Formation of disordered structures (eg. inclusion bodies *in vivo* and folding aggregates generates upon refolding of denaturant-unfolded protein at high concentration)

Protein aggregation proceeds either through the hydrophobic aggregation of the unfolded or denatured state or polymerization of the native-like conformation (Fink, 1998). Recently a series of diseases such as Alzheimer's, Parkinson's, systemic diseases originates due to misfolding of functional proteins like Aβ peptides, α-synuclein, transthyretin. We will discuss mainly the general and specific aspects of amyloid formation for various systems including its structural insight.

In Vivo

Studies that are in vivo (Latin for "within the living"; often not italicized in English) are those in which the effects of various biological entities are tested on whole, living organisms, usually animals, including humans, and plants as opposed to a partial or dead organism. This is not to be confused with experiments done in vitro ("within the glass"), i.e., in a laboratory environment using test tubes, petri dishes, etc. Examples of investigations *in vivo* include: the pathogenesis of disease by comparing the effects of bacterial infection with the effects of purified bacterial toxins; the development of antibiotics, antiviral drugs, and new drugs generally; and new surgical procedures. Consequently, animal testing and clinical trials are major elements of *in vivo* research. *In vivo* testing is often employed over *in vitro* because it is better suited for observing the overall effects of an experiment on a living subject. In drug discovery, for example, verification of efficacy *in vivo* is crucial, because *in vitro* assays can sometimes yield misleading results with drug candidate molecules that are irrelevant *in vivo* (e.g., because such molecules cannot reach their site of *in vivo* action, for example as a result of rapid catabolism in the liver).

The English microbiologist Professor Harry Smith and his colleagues in the mid-1950s showed the importance of *in vivo* studies. They found that sterile filtrates of serum from animals infected with *Bacillus anthracis* were lethal for other animals, whereas extracts of culture fluid from the same organism grown *in vitro* were not. This discovery of anthrax toxin through the use of *in vivo* experiments had a major impact on studies of the pathogenesis of infectious disease.

The maxim *in vivo veritas* ("in a living thing [there is] truth") is used to describe this type of testing and is a play on *in vino veritas*, ("in wine [there is] truth"), a well-known proverb.

In Vivo vs. ex Vivo Research

In microbiology, *in vivo* is often used to refer to experimentation done in live isolated cells rather than in a whole organism, for example, cultured cells derived from biopsies. In this situation, the more specific term is *ex vivo*. Once cells are disrupted and individual parts are tested or analyzed, this is known as *in vitro*.

Methods of use

According to Christopher Lipinski and Andrew Hopkins, "Whether the aim is to discover drugs or to gain knowledge of biological systems, the nature and properties of a chemical tool cannot be considered independently of the system it is to be tested in. Compounds that bind to isolated recombinant proteins are one thing; chemical tools that can perturb cell function another; and pharmacological agents that can be tolerated by a live organism and perturb its systems are yet another. If it were simple to ascertain the properties required to develop a lead discovered *in vitro* to one that is active *in vivo*, drug discovery would be as reliable as drug manufacturing."

In Vitro

Cloned Plants in Vitro

In vitro studies are performed with microorganisms, cells, or biological molecules outside their normal biological context. Colloquially called "test-tube experiments", these studies in biology and its subdisciplines have traditionally been done in test tubes, flasks, Petri dishes, etc., and since the onset of molecular biology, involve techniques such as the omics. Studies conducted using components of an organism that have been isolated from their usual biological surroundings permit a more detailed or more convenient analysis than can be done with whole organisms. In contrast, *in vivo* studies are those conducted in animals, including humans, and whole plants.

Definition

In vitro (Latin: *in glass*; often not italicized in English) studies are conducted using components of an organism that have been isolated from their usual biological surroundings, such as microorganisms, cells, or biological molecules. For example, microrganisms or cells can be studied in artificial culture media, and proteins can be examined in solutions. Colloquially called "test-tube experiments", these studies in biology, medicine, and their subdisciplines are traditionally done in test tubes, flasks, Petri dishes, etc. They now involve the full range of techniques used in molecular biology, such as the omics.

In contrast, studies conducted in living beings (microorganisms, animals, humans, or whole plants) are called *in vivo* .

Examples

Examples of *in vitro* studies include: the isolation, growth and identification of cells derived from multicellular organisms in (cell or tissue culture); subcellular components (e.g. mitochondria or ribosomes); cellular or subcellular extracts (e.g. wheat germ or reticulocyte extracts); purified molecules such as proteins, DNA, or RNA); and the commercial production of antibiotics and other pharmaceutical products. Viruses, which only replicate in living cells, are studied in the laboratory in cell or tissue culture, and many animal virologists refer to such work as being *in vitro* to distinguish it from *in vivo* work in whole animals.

- Polymerase chain reaction is a method for selective replication of specific DNA and RNA sequences in the test tube.

- Protein purification involves the isolation of a specific protein of interest from a complex mixture of proteins, often obtained from homogenized cells or tissues.

- *In vitro* fertilization is used to allow spermatozoa to fertilize eggs in a culture dish before implanting the resulting embryo or embryos into the uterus of the prospective mother.

- *In vitro* diagnostics refers to a wide range of medical and veterinary laboratory tests that are used to diagnose diseases and monitor the clinical status of patients using samples of blood, cells, or other tissues obtained from a patient.

- *In vitro* testing has been used to characterize specific adsorption, distribution, metabolism, and excretion processes of drugs or general chemicals inside a living organism; for example, Caco-2 cell experiments can be performed to estimate the absorption of compounds through the lining of the gastrointestinal tract; The partitioning of the compounds between organs can be determined to study distribution mechanisms; Suspension or plated cultures of primary hepatocytes or hepatocyte-like cell lines (HepG2, HepaRG) can be used to study and quantify metabolism of chemicals. These ADME process parameters can then be integrated into so called "physiologically based pharmacokinetic models" or PBPK.

Advantages

In vitro studies permit a species-specific, simpler, more convenient, and more detailed analysis than can be done with the whole organism. Just as studies in whole animals more and more replace human trials, so are *in vitro* studies replacing studies in whole animals.

Simplicity

Living organisms are extremely complex functional systems that are made up of, at a minimum, many tens of thousands of genes, protein molecules, RNA molecules, small organic compounds, inorganic ions, and complexes in an environment that is spatially organized by membranes, and in the case of multicellular organisms, organ systems. These myriad components interact with each other and with their environment in a way that processes food, removes waste, moves components to the correct location, and is responsive to signalling molecules, other organisms, light, sound, heat, taste, touch, and balance.

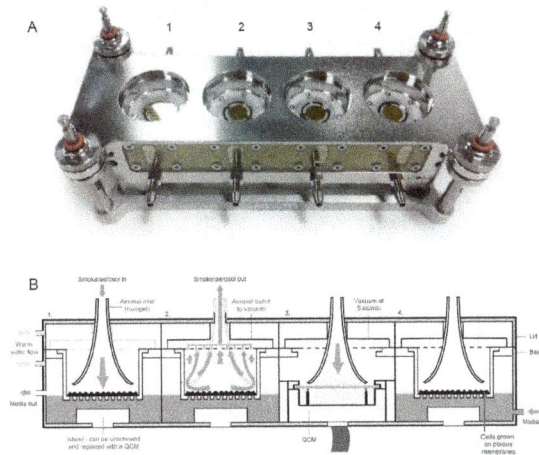

Top view of a Vitrocell mammalian exposure module "smoking robot", (lid removed) view of four separated wells for cell culture inserts to be exposed to tobacco smoke or an aerosol for an in vitro study of the effects

This complexity makes it difficult to identify the interactions between individual components and to explore their basic biological functions. *In vitro* work simplifies the system under study, so the investigator can focus on a small number of components.

For example, the identity of proteins of the immune system (e.g. antibodies), and the mechanism by which they recognize and bind to foreign antigens would remain very obscure if not for the extensive use of *in vitro* work to isolate the proteins, identify the cells and genes that produce them, study the physical properties of their interaction with antigens, and identify how those interactions lead to cellular signals that activate other components of the immune system.

Species Specificity

Another advantage of *in vitro* methods is that human cells can be studied without "extrapolation" from an experimental animal's cellular response.

Convenience, Automation

In vitro methods can be miniaturized and automated, yielding high-throughput screening methods for testing molecules in pharmacology or toxicology.

Disadvantages

The primary disadvantage of *in vitro* experimental studies is that it may be challenging to extrapolate from the results of *in vitro* work back to the biology of the intact organism. Investigators doing *in vitro* work must be careful to avoid over-interpretation of their results, which can lead to erroneous conclusions about organismal and systems biology.

For example, scientists developing a new viral drug to treat an infection with a pathogenic virus (e.g. HIV-1) may find that a candidate drug functions to prevent viral replication in an *in vitro* setting (typically cell culture). However, before this drug is used in the clinic, it must progress through a series of *in vivo* trials to determine if it is safe and effective in intact organisms (typically small animals, primates, and humans in succession). Typically, most candidate drugs that are effective *in vitro* prove to be ineffective *in vivo* because of issues associated with delivery of the drug to the affected tissues, toxicity towards essential parts of the organism that were not represented in the initial *in vitro* studies, or other issues.

In Vitro to in Vivo Extrapolation

Results obtained from *in vitro* experiments cannot usually be transposed, as is, to predict the reaction of an entire organism *in vivo*. Building a consistent and reliable extrapolation procedure from *in vitro* results to *in vivo* is therefore extremely important. Solutions include:

- Increasing the complexity of *in vitro* systems to reproduce tissues and interactions between them (as in "human on chip" systems).

- Using mathematical modeling to numerically simulate the behavior of the complex system, where the *in vitro* data provide model parameter values.

These two approaches are not incompatible; better *in vitro* systems provide better data to mathematical models. However, increasingly sophisticated *in vitro* experiments collect increasingly numerous, complex, and challenging data to integrate. Mathematical models, such as systems biology models, are much needed here.

Extrapolating in Pharmacology

In pharmacology, IVIVE can be used to approximate pharmacokinetics (PK) or pharmacodynamics (PD). Since the timing and intensity of effects on a given target depend on the concentration time course of candidate drug (parent molecule or metabolites) at that target site, *in vivo* tissue and organ sensitivities can be completely different or even inverse of those observed on cells cultured and exposed *in vitro*. That indicates that extrapolating effects observed *in vitro* needs a quantitative model of *in vivo* PK. Physiologically based PK (PBPK) models are generally accepted to be central to the extrapolations.

In the case of early effects or those without intercellular communications, the same cellular exposure concentration is assumed to cause the same effects, both qualitatively and quantitatively, *in vitro* and *in vivo*. In these conditions, developing a simple PD model of the dose–response relationship observed *in vitro*, and transposing it without changes to predict *in vivo* effects is not enough.

Amyloid Formation, Molecular Basis and Protein Misfolding in Cell

Proteins undergo formation of toxic deposits commonly known as fibrils (amyloid fibrils) under varying conditions which is found to be the generic feature of polypeptides (Dobson, 2001; Gazit, 2002). Amyloid deposits were first identified by Rudolph Virchow in 1854 by iodine staining of brain section. In 1959, electron microscopy studies reveal that different amyloid tissues exhibit a common structural motif in the fixed tissue section (Cohen and Calkins, 1959). Protein misfolding leads to formation of these amyloidal plaques, that occurs for both disease-related and disease-unrelated proteins (Guijarro et al, 1998). Table shows list of diseases that arise due to amyloid formation from specific proteins.

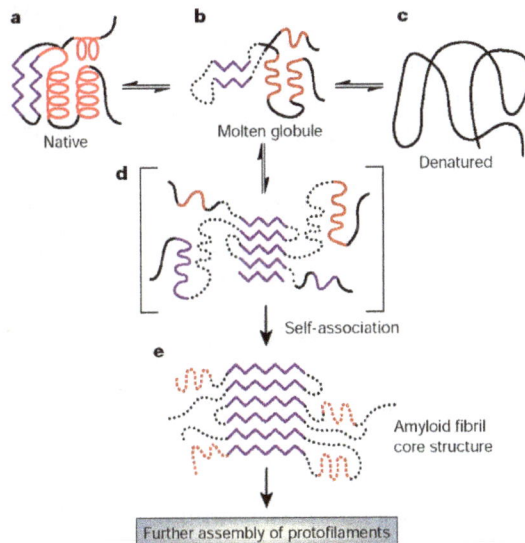

Conformational alterations takes place under certain biophysical conditions which lead to formation of partially folded metastable state that facilitates intermolecular interactions and finally form oligomeric species (Rochet and Lansbury, 2000). During this oligomeric species formation mainly the secondary structural element that is β-sheet content increases.

Protein folding in eukaryotic cells takes place either in the cytosol or in the endoplasmic reticulum. High degree of molecular crowding in the cell signifies improperly folded molecules which will either form self aggregates or aggregate with the cellular components. Folding catalysts, molecular chaperones, and several quality control mechanisms are there which assist in the maintenance of correct folding of proteins (Munoz, 2008).

Table: Amyloid deposition from different proteins, organs affected and localization of the deposits (Munoz, 2008)

Clinical syndrome	Organs affected	Cellular localization	Plaque components
Alzheimer's disease	*Brain: cerebral cortex, hippocampus*	Extracellular Tangles in neuronal cytoplasm	Amyloid β- peptide Tau protein
Parkinson's disease	*Brain: substantia nigra, hypothalamus*	Neuronal cytoplasm	α-synuclein
Polyglutamine expansion disease	*Brain: Striatum, cerebral cortex*	Neuromal nuclei and cytoplasm	Long glutamine stretches within certain proteins eg. Huntington
Spongiform encephalopathy	*Brain: cortex, thalamus*	Extra and intracellular	Prion
Type II diabetes	*Pancreas*	Extracellular	Islet amyloid polypeptide
Familial amyloidotic polyneuropathy	*Systemic; peripheral nerves, heart*	Extracellular	Mutant transthyretin and fragments thereof
Senile systemic amyloidosis	*Systemic, cardiace*	Extracellular	Wild-type transthyretin and fragments thereof
Haemodialysis related amyloidosis	*Systemic; joints, bones*	Extracellular	β2-microglobulin
Finnish hereditary amyloidosis	*Systemic; ocular*	Extracellular	Fragments of mutant gelsolin
Hereditary systemic amyloidosis	*Systemic: renal and visceral disease*	Extracellular	Mutant lysozyme

Conditions Facilitating Amyloid Formation in Vivo and in Vitro

In vivo amyloid formation is associated with enhanced level of amyloid forming proteins due to any factor which regulates the equilibrium between correctly and partially folded molecules. Protein mutations, environmental changes and also chemical modifications are known to promote amyloid formation.

Mutation either reduces conformational stability of proteins or increases the population of the amyloid prone polypeptide sequences which ultimately favors amyloid formation. It has been also found that interaction with other cellular components may affect amyloid formation. Proteoglycans were found to accelerates and stabilize fibrils and thus play crucial role in the pathology of amyloid formation (Sipe, 1992). Metal ions specifically Cu(II), Fe(II), Zn(II) were found to have unique property of either accelerating or inhibiting fibrillation (Munoz, 2008). Post translational modification of proteins also found to play a critical role in amyloid formation. For example amyloid formation of two intrinsically disordered proteins is favored due to phosphorylation. In addition to this destruction of quality control machinery results in the formation of misfolded intermediates which favors amyloid formation.

In vitro amyloid formation can be accelerated by varying several external factors such as change in pH, elevated temperature, varying salt or denaturant concentrations or solvents. In these conditions tertiary interaction within a protein structure is destabilized but secondary interaction is favored. Though these conditions are not applicable for in vivo amyloid formation but they can be used to understand underlying molecular mechanism.

Amyloid Formation of Different Systems: Disease Related and Disease Unrelated Proteins

Alzheimer's Disease (AD) and Tau Related Disease

AD pathogenesis results in the progressive loss in memory, acute recognition disorder in identifying person, objects etc. It is associated with degeneration of neurons and changed neuronal connections. AD involves formation of two types of aggregates: extracellular and intracellular. Extracellular and intracellular aggregates usually originate from Aβ peptides and microtubule associated protein tau respectively. Genetic mutation is considered to be the major reason behind these pathological disorders (Ross and Poirier).

Parkinson's Disease (PD)

The symptoms of PD are latent shivering, rigidity, slowness in movements, and postural and autonomic instability. It mainly arises due to degeneration of dopaminergic neurons in the substantia nigra of the mid brain. Gene mutation has drawn our attention toward early onset of PD. The characteristic feature of adult beginning PD is the formation of Lewy body, major constituent of which is aggregated α-synuclein, near the nucleus in the cytoplasm of neurons.

Protein Aggregation in General: Human Serum Albumin (HSA) and Hen Egg White Lysozyme (HEWL)

HSA, the most abundant plasma protein, and HEWL (structural homologue of human lysozyme, origin of systemic amyloidosis disease) are two disease unrelated large globular proteins. Though, they are considered to be suitable model *in vitro* for the better understanding of underlying molecular mechanism in general. Aggregation of HSA (Juarez et al., 2009; Pandey et al., 2010) and HEWL (Xie et al., 2012; Ghosh et al., 2012) was found to be greatly affected in the presence of varying external conditions such as change in pH, temperature, metal ions etc. Reports have also revealed different kinetic modes of self assembly. The nature of the aggregates, morphological diversity, different amyloid specific dye such as Congo Red and Thioflavin T (ThT) binding property suggest that aggregation of proteins of different sequence and nature are able to form amyloidal aggregates of similar basic property.

Basic Idea About Amyloid Fibrillar Architecture

Structural complexity of amyloid fibrils (insoluble and non crystalline) has created severe difficulties in studying its structural details both in vivo and in vitro. Even techniques such as solution NMR and single XRD were also found to be difficult to apply. Thus observations from different studies have been compiled sequentially to get a picture of structure of fibrils. Cross β-sheet motif is found to be the common structural unit of all the amyloid fibrillar structure where β-sheets are hydrogen bonded along the length of the fibrils and β-strands run perpendicular to the long axis of fibrils (Munoz, 2008).

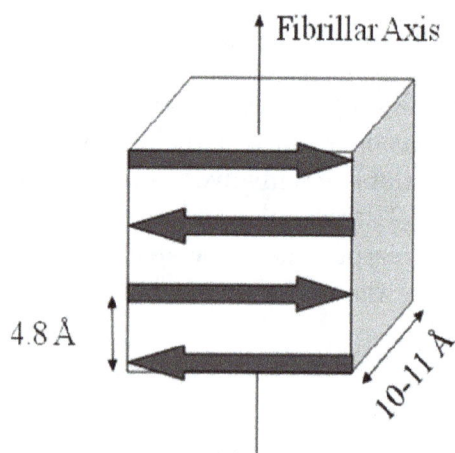

Protein aggregation is an emerging field of research which has gained much of our interest in recent times due to the devastating effect of neurological disorders such as Alzheimer's, Parkinson's over a large popularity worldwide. Thus diagnosis and therapeutic approaches toward these diseases become immensely important. Here in this chapter, we have described different biophysical methods of characterizing amyloid formation.

Biophysical techniques used for characterizing amyloid formation are as follows:

- UV-vis spectroscopy

- Fluorescence spectroscopy

- Circular dichroism spectroscopy

- FTIR measurement

- Fluorescence microscopy

- Transmission electron microscopy

UV-vis Spectroscopy: Congo Red Binding Assay

Congo red is considered to be an amyloid marker dye since past years (Klunk et al., 1989). Congo red absorbs nears ~498 nm. Binding with amyloid fibrils causes a red shift of ~45-50 nm in the absorbance spectra. Thus, from this observation, it is possible to perform the initial characterization of amyloid fibrils. Congo red is also used to stain amyloid fibrils and obtain images using fluorescence microscopy.

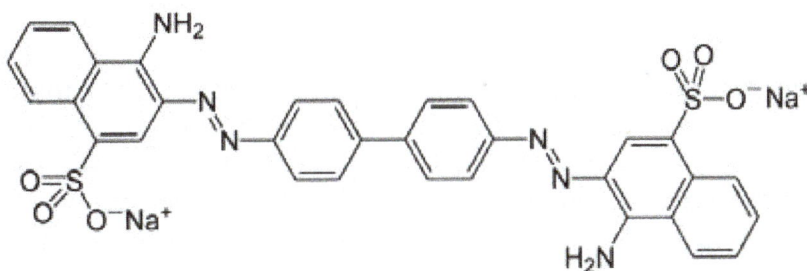

Congo Red

Fluorescence Spectroscopy

Thioflavin T (ThT) Binding Study using Fluorescence Spectroscopy

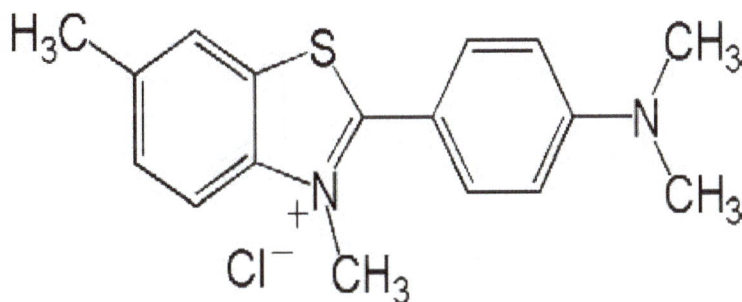

Thioflavin T (ThT)

Thioflavin T (ThT) is an amyloid specific dye which has unique property to bind with amyloid fibrils and exhibits intense fluorescence at ~480-490 nm upon excitation at

450 nm. This is the most useful technique to characterize amyloid formation. ThT binding affinity of amyloid fibrils has enabled researchers to determine the nature of amyloid fibrillation pathway. Previous studies have revealed that using the ThT binding affinity of protein aggregates, the nature of the complex amyloid formation pathway can be explored. Thus, the ThT kinetic data (ThT fluorescence intensities at different interval of time) for were subjected to a nonlinear least squares curve- fitting to a stretched exponential function:

$$F = F\infty + \Delta F \, \exp\left[-\left(ksp^{*}t\right)^{n}\right]$$

where F, $F\infty$ and ΔF are the observed fluorescence intensity at time t, final fluorescence intensity and fluorescence amplitude respectively and ksp represents the rate of spontaneous fibril formation. When value of n lies in between 0 and 1, it indicates kinetics with multiple exponential functions that signifies numerous events linked with fibrillation. However when value of n > 1, a sigmoidal transition is expected to occur with an initial lag phase which involves different intermediate species (Morozova–Roche et al, 1999; Jund et al, 2001; Hamada and Dobson, 2002).

Intrinsic (Trp) Fluorescence Study

Trp is considered to be an intrinsic fluorophore which is very sensitive to any change in the microenvironment of proteins. Thus one can monitor Trp fluorescence intensity of proteins over at definite interval of time and any change in the intensity or shift in the spectra will denote corresponding change in the system. Usually, to monitor the Trp fluorescence, protein solutions excited at 295 nm (to avoid any contribution from Tyr residues, excitation at 280 nm is avoided). During fibrillation tertiary structure of the proteins is altered, and it is likely that the environment of Trp residues also changes. Thus, monitoring Trp fluorescence to get an insight about the mechanistic aspect of protein aggregation has become a very useful technique.

ANS Binding Assay

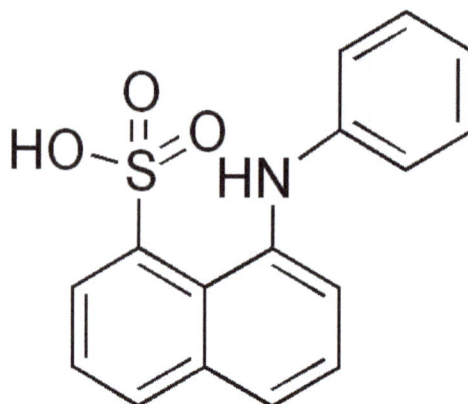

8-anilino-1-naphthalenesulfonic acid (ANS)

8-anilino-1-naphthalenesulfonic acid (ANS) is widely used as a hydrophobic marker dye to identify the hydrophobic region of proteins (Semisotnov et al., 1991). Protein aggregation is associated with formation of large aggregates where noncovalent (exposure of hydrophobic region) interactions take place. Thus, ANS binding assay before and after protein aggregation provides a clear picture of hydrophobic exposure during aggregation. Free ANS emits at ~490-510 nm upon excitation at 370 nm. Binding with the hydrophobic region of protein causes a blue shift in its fluorescence spectrum and thus λmax appears at ~470-475 nm. Therefore, it appears that monitoring ANS binding affinity of the protein aggregates so formed one can speculate whether hydrophobic exposure has taken place or not.

Circular Dichroism Spectroscopy

Proteins undergo a significant conformational alteration during amyloid formation that is α-helix to β-sheet transformation takes place. Circular dichroism spectroscopy is a useful technique to monitor this conformational change of proteins (Kirkitadze et al., 2001). Origin of CD of proteins is due to the induced rotational strength in transitions of amide bonds which attributed due to the structural asymmetry of proteins. Far UV-CD spectra of native structured proteins (containing both α-helix and β-sheet) give two minima, one at ~208 nm ($\pi \rightarrow \pi^*$ transition) and another one at 222 nm ($n \rightarrow \pi^*$ transition) characteristic of α-helical content. There appears one negative band and one positive band at ~216-218 nm and 195 nm respectively for proteins containing β-sheet structure. These characteristic signals obtained in circular dichroism spectroscopy enable one to monitor the change in secondary structural content of proteins during fibrillation. Near UV-CD spectra of proteins give characteristic signals at ~262, ~268 nm (correspond to the disulfide bonds) and at ~275 nm,~285 nm (which and aromatic amino acid residues) respectively. Thus any change in the near UV-CD spectra of proteins during aggregation indicates tertiary structural alterations of proteins due to aggregation.

FTIR Spectroscopy

FTIR spectroscopy is basically a method to determine the absorption of IR radiation of a sample in terms of wavelength and intensity. Polypeptide backbone and repeating units of proteins absorb IR radiation and the two major bands arise are amide I and amide II bands respectively. Amide I band (1700-1600 cm^{-1}) and Amide II band (1575-1480 cm^{-1}) arise mainly due to C=O stretching vibration of the peptide linkages and in plane NH bending respectively (Susi and Byler, 1986; Kong and Yu, 2007). Amide I band is extremely sensitive to any little changes in the molecular geometry and hydrogen bonding pattern which enables it to be useful for protein secondary structural analysis (Susi and Byler, 1986). Band arises in between 1660-1650 cm^{-1} corresponds to the α-helix structure. The β-sheet structure corresponds to the bands arise in the region of 1640-1620 cm^{-1} and 1695-1690 cm^{-1}. Bands appear around 1670, 1683, 1688 and 1694 cm^{-1} correspond to β-turn and in the region of 1648-1640 cm^{-1} correspond to random coil structure (Krimm and Bandekar, 1986). Thus, similar to circular dichroism spectroscopy, exploiting FTIR spectroscopy one can unravel the conformational alteration of protein structure during fibrillation as it is associated with α-helix to β-sheet transformation. So, comparing the band appears in the native protein and its aggregated form one can easily conclude the secondary structural changes of protein during fibrillation.

Fluorescence Microscopy

ThT is extensively used to obtain microscopic images via fluorescence microscopy. The strategy behind this technique is similar. Protein samples in the aggregated form which are expected to show fibrillar structures stained using ThT dye, air dried. Then they were subjected to fluorescence using excitation wavelength ~450 nm and the images obtained exhibiting green fluorescence (if they are amyloidal in nature) were captured. Thus, presence of amyloidal aggregates can be determined via fluorescence microscopy using ThT as marker dye.

Transmission Electron Microscopy (TEM)

Transmission electron microscopy (TEM) is a widely used technique to explore the fine structure of the ultra thin specimen. In this technique electron beam passes through the specimen and interact, resulting in an image formation which is magnified and focused followed by capturing via a CCD camera. TEM is an extremely useful technique which is very useful for determining morphological aspect of biological substances such as protein, DNA etc. Protein aggregation results amyloid formation which generates various morphology such as threadlike, rodlike, hairlike, ribbonlike, wormlike etc which can be easily viewed through TEM experiments. Morphological diversity of protein aggregates also depends on the protein chosen and conditions applied. All these conditions are essentially play critical role and exert notable effects in terms of morphological variations of protein aggregates which can be explored using TEM.

Protein Isolation and Characterization - I

Proteins are quite diverse in nature. Protein expression is strongly regulated for normal functioning of a cell or organism. To be able to understand protein structure and function in detail, they often need to be separated from several cellular components like lipids, nucleic acids, sugars, etc. and isolated to homogeneity. A given cell may have thousands of proteins. It is quite challenging to isolate a particular one from the huge number of proteins. Also after obtaining a protein to near homogeneity one has to ensure that it retains its native biological characteristics of structure as well as activity. Thus its purification and characterization, enzymatic activity and structural elucidation are essential for a complete understanding of the protein.

Proteins can be purified by exploiting specific properties that include solubility, size, charge, and binding affinity towards some specific agents. It must be remembered that protein purification strictly depends upon the precise nature of the protein.

General methods include

- Precipitation

- Extraction and

- Chromatographic separation

Protein Precipitation

The most common source of proteins is microbial cells or tissue. Cytosolic proteins are highly water soluble and their solubility is a function of the ionic strength and pH of the solution. The most useful method of precipitating protein from its solution is to add salt to it. The commonly used salt for this purpose is ammonium sulfate, due to its high solubility even at lower temperatures. Both ions of ammonium sulfate are high in the Hofmeister series or lyotropic series which is a classification of ions in order of their ability to salt out or salt in proteins. The earliest step in any protein purification procedure needs to rupture cells; to release their proteins into a solution, often called a crude extract or lysate. Sometimes differential centrifugation may need to collects a specific sub cellular part or organelles.

Salting Out

Most proteins are less soluble at high salt concentrations, an effect called salting out. Proteins in aqueous solutions are highly hydrated, and with the addition of salt, the water molecules become more attracted to the salt than to the protein due to the higher charge. The addition of ammonium sulfate in the requisite amount can selectively precipitate a protein of interest while others remain in solution. For example, 0.8 M ammonium sulfate can precipitate fibrinogen, whereas a concentration of 2.4 M is needed to precipitate serum albumin. Salting out is very useful for concentrating dilute solutions of proteins, and has no adverse structural effects.

In this processes, the ammonium sulfate concentration is increased stepwise by adding solid ammonium sulfate in small quantities. The amount depends on the volume of the solution and the percentage saturation of the salt needed. It is possible to calculate how much is needed from available published nomograms. For protein purification, a step precipitation is carried out in which the precipitated protein is removed by centrifugation and the ammonium sulfate concentration increased to a value that will precipitate most of the protein of interest. In most cases this leaves out the protein contaminants in solution. The precipitated protein of interest is further recovered by the process of centrifugation and the pellet obtained dissolved in fresh buffer to be purified further. If required the above process is repeated usually with a different concentration of ammonium sulfate for precipitation of a different protein from the supernatant.

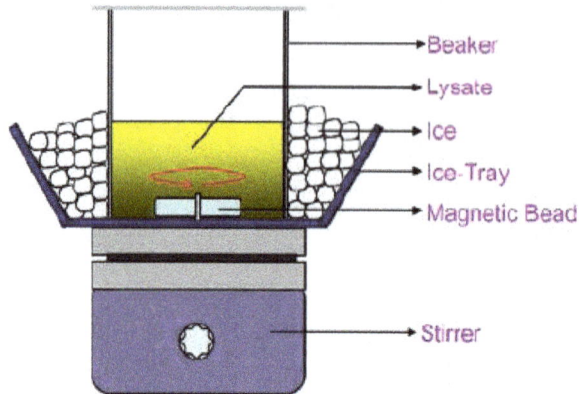

Protein precipitation using ammonium sulfate (Needs to be drawn later)

Salting In

As mentioned earlier, the solubility of proteins is strongly dependent on the salt concentration or ionic strength of the medium. Proteins are usually poorly soluble in pure water. Their solubility increases as the ionic strength increases, because more and more of the well-hydrated inorganic ions (blue circles) are bound to the protein's surface, preventing aggregation of the molecules (salting in).

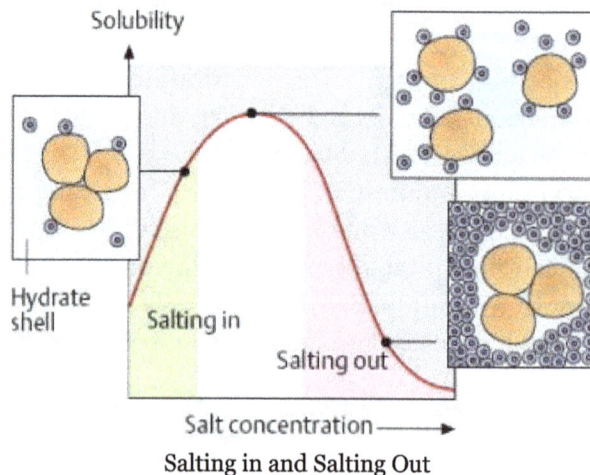

Salting in and Salting Out

Dialysis

A solution containing the protein of interest must be further altered before purification steps are possible. Proteins can be separated from small molecules by taking advantage of the larger size of the protein compared to other molecules. Dialysis through a semi permeable membrane (SEM), such as a cellulose membrane with well defined pores is often done to free a protein of choice from other contaminants. The partially purified protein solution is placed in a dialysis bag and the bag suspended in a much larger volume of buffer of appropriate ionic strength.

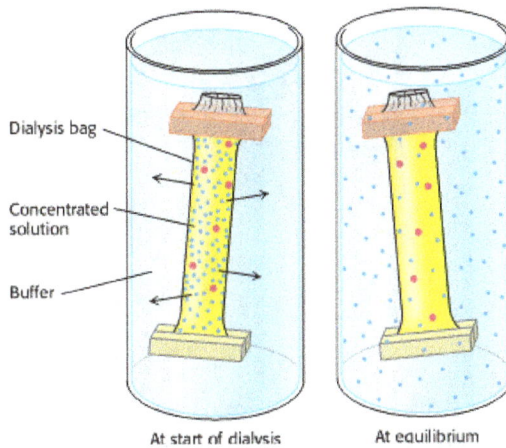

Dialysis bag

Concentrated solution

Buffer

At start of dialysis   At equilibrium

Protein molecules (red) are retained within the dialysis bag, whereas small molecules (blue) diffuse into the surrounding medium

Proteins molecules having significantly greater dimensions than the pore diameter of the dialysis bag, whereas smaller molecules and ions cross the pores of such membranes and emerge in the dialysate outside the bag. This technique is useful for removing a salt or other small molecule, but it will not distinguish between proteins effectively. Dialysis bags of definite molecular weight cutoff are often employed to purify protein of definite size. For example, a dialysis bag having a 12,000 molecular weight cutoff can be safely used to purify proteins having molecular weights of ~20,000 but may not be appropriate for proteins with a molecular weight of 10,000.

Chromatography

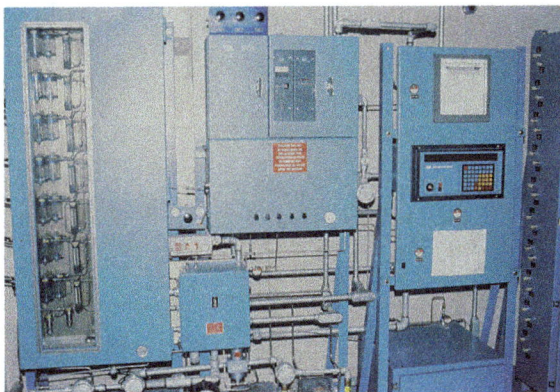

Pictured is a sophisticated gas chromatography system. This instrument records concentrations of acrylonitrile in the air at various points throughout the chemical laboratory.

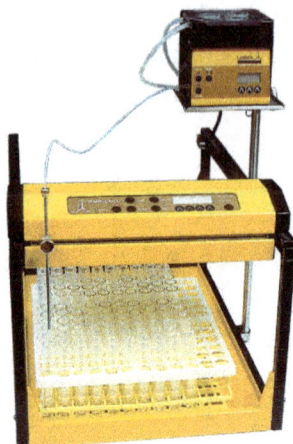

Automated fraction collector and sampler for chromatographic techniques

Chromatography is a laboratory technique for the separation of a mixture. The mixture is dissolved in a fluid called the mobile phase, which carries it through a structure holding another material called the stationary phase. The various constituents of the mixture travel at different speeds, causing them to separate. The separation is based on differential partitioning between the mobile and stationary phases. Subtle differences in a compound's partition coefficient result in differential retention on the stationary phase and thus changing the separation.

Chromatography may be preparative or analytical. The purpose of preparative chromatography is to separate the components of a mixture for later use, and is thus a form of purification. Analytical chromatography is done normally with smaller amounts of material and is for establishing the presence or measuring the relative proportions of analytes in a mixture. The two are not mutually exclusive.

History

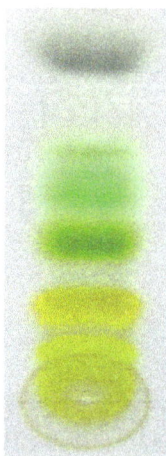

Thin layer chromatography is used to separate components of a plant extract, illustrating the experiment with plant pigments that gave chromatography its name

Chromatography was first employed in Russia by the Italian-born scientist Mikhail Tsvet in 1900. He continued to work with chromatography in the first decade of the 20th century, primarily for the separation of plant pigments such as chlorophyll, carotenes, and xanthophylls. Since these components have different colors (green, orange, and yellow, respectively) they gave the technique its name. New types of chromatography developed during the 1930s and 1940s made the technique useful for many separation processes.

Chromatography technique developed substantially as a result of the work of Archer John Porter Martin and Richard Laurence Millington Synge during the 1940s and 1950s, for which they won a Nobel prize. They established the principles and basic techniques of partition chromatography, and their work encouraged the rapid development of several chromatographic methods: paper chromatography, gas chromatography, and what would become known as high performance liquid chromatography. Since then, the technology has advanced rapidly. Researchers found that the main principles of Tsvet's chromatography could be applied in many different ways, resulting in the different varieties of chromatography described below. Advances are continually improving the technical performance of chromatography, allowing the separation of increasingly similar molecules.

Chromatography Terms

- The analyte is the substance to be separated during chromatography. It is also normally what is needed from the mixture.

- Analytical chromatography is used to determine the existence and possibly also the concentration of analyte(s) in a sample.

- A bonded phase is a stationary phase that is covalently bonded to the support particles or to the inside wall of the column tubing.

- A chromatogram is the visual output of the chromatograph. In the case of an optimal separation, different peaks or patterns on the chromatogram correspond to different components of the separated mixture.

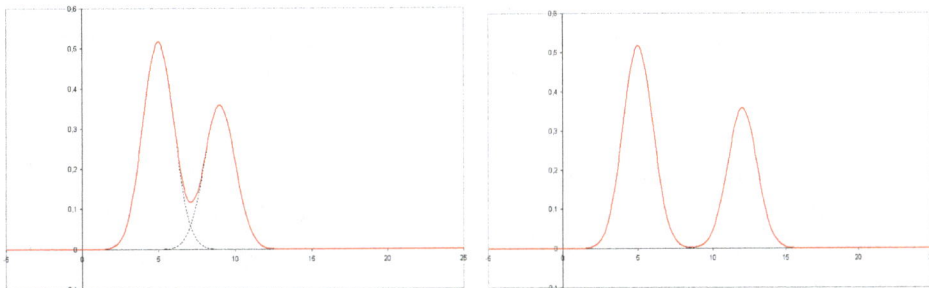

Plotted on the x-axis is the retention time and plotted on the y-axis a signal (for example obtained by a spectrophotometer, mass spectrometer or a variety of other detectors) corresponding to the response created by the analytes exiting the system. In the case of an optimal system the signal is proportional to the concentration of the specific analyte separated.

- A chromatograph is equipment that enables a sophisticated separation, e.g. gas chromatographic or liquid chromatographic separation.

- Chromatography is a physical method of separation that distributes components to separate between two phases, one stationary (stationary phase), the other (the mobile phase) moving in a definite direction.

- The eluate is the mobile phase leaving the column. This is also called effluent.

- The eluent is the solvent that carries the analyte.

- The eluite is the analyte, the eluted solute.

- An eluotropic series is a list of solvents ranked according to their eluting power.

- An immobilized phase is a stationary phase that is immobilized on the support particles, or on the inner wall of the column tubing.

- The mobile phase is the phase that moves in a definite direction. It may be a liquid (LC and Capillary Electrochromatography (CEC)), a gas (GC), or a supercritical fluid (supercritical-fluid chromatography, SFC). The mobile phase consists of the sample being separated/analyzed and the solvent that moves the sample through the column. In the case of HPLC the mobile phase consists of a non-polar solvent(s) such as hexane in normal phase or a polar solvent such as methanol in reverse phase chromatography and the sample being separated. The mobile phase moves through the chromatography column (the stationary phase) where the sample interacts with the stationary phase and is separated.

- Preparative chromatography is used to purify sufficient quantities of a substance for further use, rather than analysis.

- The retention time is the characteristic time it takes for a particular analyte to pass through the system (from the column inlet to the detector) under set conditions.

- The sample is the matter analyzed in chromatography. It may consist of a single component or it may be a mixture of components. When the sample is treated in the course of an analysis, the phase or the phases containing the analytes of interest is/are referred to as the sample whereas everything out of interest separated from the sample before or in the course of the analysis is referred to as waste.

- The solute refers to the sample components in partition chromatography.

- The solvent refers to any substance capable of solubilizing another substance, and especially the liquid mobile phase in liquid chromatography.

- The stationary phase is the substance fixed in place for the chromatography procedure. Examples include the silica layer in thin layer chromatography.

- The detector refers to the instrument used for qualitative and quantitative detection of analytes after separation.

Chromatography is based on the concept of partition coefficient. Any solute partitions between two immiscible solvents. When we make one solvent immobile (by adsorption on a solid support matrix) and another mobile it results in most common applications of chromatography. If the matrix support, or stationary phase, is polar (e.g. paper, silica etc.) it is forward phase chromatography, and if it is non-polar (C-18) it is reverse phase.

Techniques by Chromatographic Bed Shape

Column Chromatography

Column chromatography is a separation technique in which the stationary bed is within a tube. The particles of the solid stationary phase or the support coated with a liquid stationary phase may fill the whole inside volume of the tube (packed column) or be concentrated on or along the inside tube wall leaving an open, unrestricted path for the mobile phase in the middle part of the tube (open tubular column). Differences in rates of movement through the medium are calculated to different retention times of the sample.

In 1978, W. Clark Still introduced a modified version of column chromatography called flash column chromatography (flash). The technique is very similar to the traditional column chromatography, except for that the solvent is driven through the column by applying positive pressure. This allowed most separations to be performed in less than 20 minutes, with improved separations compared to the old method. Modern flash chromatography systems are sold as pre-packed plastic cartridges, and the solvent is pumped through the cartridge. Systems may also be linked with detectors and fraction collectors providing automation. The introduction of gradient pumps resulted in quicker separations and less solvent usage.

In expanded bed adsorption, a fluidized bed is used, rather than a solid phase made by a packed bed. This allows omission of initial clearing steps such as centrifugation and filtration, for culture broths or slurries of broken cells.

Phosphocellulose chromatography utilizes the binding affinity of many DNA-binding proteins for phosphocellulose. The stronger a protein's interaction with DNA, the higher the salt concentration needed to elute that protein.

Planar Chromatography

Planar chromatography is a separation technique in which the stationary phase is present as or on a plane. The plane can be a paper, serving as such or impregnated by a substance as the stationary bed (paper chromatography) or a layer of solid particles spread on a support such as a glass plate (thin layer chromatography). Different compounds in the sample mixture travel different distances according to how strongly they interact with the stationary phase as compared to the mobile phase. The specific Retention factor (R_f) of each chemical can be used to aid in the identification of an unknown substance.

Paper Chromatography

Paper chromatography is a technique that involves placing a small dot or line of sample solution onto a strip of *chromatography paper*. The paper is placed in a container with a shallow layer of solvent and sealed. As the solvent rises through the paper, it meets the sample mixture, which starts to travel up the paper with the solvent. This paper is made of cellulose, a polar substance, and the compounds within the mixture travel farther if they are non-polar. More polar substances bond with the cellulose paper more quickly, and therefore do not travel as far.

Thin Layer Chromatography (TLC)

Thin layer chromatography (TLC) is a widely employed laboratory technique use to separate different biochemicals on the basis of their size and is similar to paper chromatography. However, instead of using a stationary phase of paper, it involves a stationary phase of a thin layer of adsorbent like silica gel, alumina, or cellulose on a flat, inert substrate. Compared to paper, it has the advantage of faster runs, better separations, and the choice between different adsorbents. For even better resolution and to allow for quantification, high-performance TLC can be used. An older popular use had been to differentiate chromosomes by observing distance in gel (separation of was a separate step).

Displacement Chromatography

The basic principle of displacement chromatography is: A molecule with a high affinity for the chromatography matrix (the displacer) competes effectively for binding sites, and thus displace all molecules with lesser affinities. There are distinct differences between displacement and elution chromatography. In elution mode, substances typically emerge from a column in narrow, Gaussian peaks. Wide separation of peaks, preferably to baseline, is desired for maximum purification. The speed at which any component of a mixture travels down the column in elution mode depends on many factors. But for two substances to travel at different speeds, and thereby be resolved, there must be substantial differences in some interaction between the biomolecules and the chromatography matrix. Operating parameters are adjusted to maximize the effect of this difference. In many cases, baseline separation of the peaks can be achieved only with gradient elution and low column loadings. Thus, two drawbacks to elution mode chromatography, especially at the preparative scale, are operational complexity, due to gradient solvent pumping, and low throughput, due to low column loadings. Displacement chromatography has advantages over elution chromatography in that components are resolved into consecutive zones of pure substances rather than "peaks". Because the process takes advantage of the nonlinearity of the isotherms, a larger column feed can be separated on a given column with the purified components recovered at significantly higher concentrations.

Techniques by Physical State of Mobile Phase

Gas Chromatography

Gas chromatography (GC), also sometimes known as gas-liquid chromatography, (GLC), is a separation technique in which the mobile phase is a gas. Gas chromatographic separation is always carried out in a column, which is typically "packed" or "capillary". Packed columns are the routine work horses of gas chromatography, being cheaper and easier to use and often giving adequate performance. Capillary columns generally give far superior resolution and although more expensive are becoming widely used, especially for complex mixtures. Both types of column are made from non-adsorbent and chemically inert materials. Stainless steel and glass are the usual materials for packed columns and quartz or fused silica for capillary columns.

Gas chromatography is based on a partition equilibrium of analyte between a solid or viscous liquid stationary phase (often a liquid silicone-based material) and a mobile gas (most often helium). The stationary phase is adhered to the inside of a small-diameter (commonly 0.53 – 0.18mm inside diameter) glass or fused-silica tube (a capillary column) or a solid matrix inside a larger metal tube (a packed column). It is widely used in analytical chemistry; though the high temperatures used in GC make it unsuitable for high molecular weight biopolymers or proteins (heat denatures them), frequently encountered in biochemistry, it is well suited for use in the petrochemical, environmental monitoring and remediation, and industrial chemical fields. It is also used extensively in chemistry research.

Liquid Chromatography

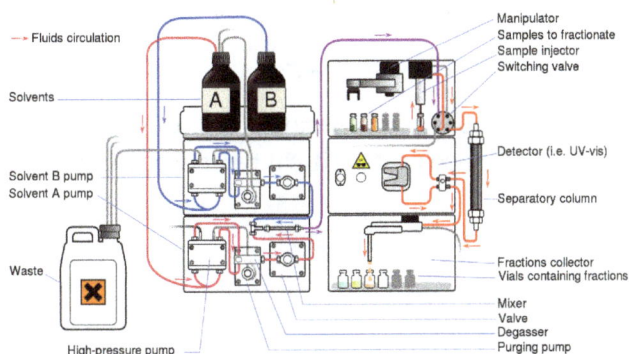

Preparative HPLC apparatus

Liquid chromatography (LC) is a separation technique in which the mobile phase is a liquid. It can be carried out either in a column or a plane. Present day liquid chromatography that generally utilizes very small packing particles and a relatively high pressure is referred to as high performance liquid chromatography (HPLC).

In HPLC the sample is forced by a liquid at high pressure (the mobile phase) through a column that is packed with a stationary phase composed of irregularly or spherically shaped particles, a porous monolithic layer, or a porous membrane. HPLC is historically divided into two different sub-classes based on the polarity of the mobile and stationary phases. Methods in which the stationary phase is more polar than the mobile phase (e.g., toluene as the mobile phase, silica as the stationary phase) are termed normal phase liquid chromatography (NPLC) and the opposite (e.g., water-methanol mixture as the mobile phase and C18 (octadecylsilyl) as the stationary phase) is termed reversed phase liquid chromatography (RPLC).

Specific techniques under this broad heading are listed below.

Affinity Chromatography

Affinity chromatography is based on selective non-covalent interaction between an analyte and specific molecules. It is very specific, but not very robust. It is often used in biochemistry in the purification of proteins bound to tags. These fusion proteins are labeled with compounds such as His-tags, biotin or antigens, which bind to the stationary phase specifically. After purification, some of these tags are usually removed and the pure protein is obtained.

Affinity chromatography often utilizes a biomolecule's affinity for a metal (Zn, Cu, Fe, etc.). Columns are often manually prepared. Traditional affinity columns are used as a preparative step to flush out unwanted biomolecules.

However, HPLC techniques exist that do utilize affinity chromatography properties. Immobilized Metal Affinity Chromatography (IMAC) is useful to separate aforementioned molecules based on the relative affinity for the metal (I.e. Dionex IMAC). Often these columns can be loaded with different metals to create a column with a targeted affinity.

Supercritical Fluid Chromatography

Supercritical fluid chromatography is a separation technique in which the mobile phase is a fluid above and relatively close to its critical temperature and pressure.

Techniques by Separation Mechanism

Ion Exchange Chromatography

Ion exchange chromatography (usually referred to as ion chromatography) uses an ion exchange mechanism to separate analytes based on their respective charges. It is usually performed in columns but can also be useful in planar mode. Ion exchange chromatography uses a charged stationary phase to separate charged compounds including anions, cations, amino acids, peptides, and proteins. In conventional methods the stationary phase is an ion exchange resin that carries charged functional groups that interact with oppositely charged groups of the compound to retain. Ion exchange chromatography is commonly used to purify proteins using FPLC.

Size-exclusion Chromatography

Size-exclusion chromatography (SEC) is also known as gel permeation chromatography (GPC) or gel filtration chromatography and separates molecules according to their size (or more accurately according to their hydrodynamic diameter or hydrodynamic volume). Smaller molecules are able to enter the pores of the media and, therefore, molecules are trapped and removed from the flow of the mobile phase. The average residence time in the pores depends upon the effective size of the analyte molecules. However, molecules that are larger than the average pore size of the packing are excluded and thus suffer essentially no retention; such species are the first to be eluted. It is generally a low-resolution chromatography technique and thus it is often reserved for the final, "polishing" step of a purification. It is also useful for determining the tertiary structure and quaternary structure of purified proteins, especially since it can be carried out under native solution conditions.

Expanded bed Adsorption Chromatographic Separation

An expanded bed chromatographic adsorption (EBA) column for a biochemical separation process comprises a pressure equalization liquid distributor having a self-cleaning function below a porous blocking sieve plate at the bottom of the expanded bed, an upper part nozzle assembly having a backflush cleaning function at the top of the expanded bed, a better distribution of the feedstock liquor added into the expanded bed ensuring that the fluid passed through the expanded bed layer displays a state of piston flow. The expanded bed layer displays a state of piston flow. The expanded bed chromatographic separation column has advantages of increasing the separation efficiency of the expanded bed.

Expanded-bed adsorption (EBA) chromatography is a convenient and effective technique for the capture of proteins directly from unclarified crude sample. In EBA chromatography, the settled bed is first expanded by upward flow of equilibration buffer. The crude feed, a mixture of soluble proteins, contaminants, cells, and cell debris, is then passed upward through the expanded bed. Target proteins are captured on the adsorbent, while particulates and contaminants pass through. A change to elution buffer while maintaining upward flow results in desorption of the target protein in expanded-bed mode. Alternatively, if the flow is reversed, the adsorbed particles will quickly settle and the proteins can be desorbed by an elution buffer. The mode used for elution (expanded-bed versus settled-bed) depends on the characteristics of the feed. After elution, the adsorbent is cleaned with a predefined cleaning-in-place (CIP) solution, with cleaning followed by either column regeneration (for further use) or storage.

Special Techniques

Reversed-phase Chromatography

Reversed-phase chromatography (RPC) is any liquid chromatography procedure in which the mobile phase is significantly more polar than the stationary phase. It is so named because in normal-phase liquid chromatography, the mobile phase is significantly less polar than the stationary phase. Hydrophobic molecules in the mobile phase tend to adsorb to the relatively hydrophobic stationary phase. Hydrophilic molecules in the mobile phase will tend to elute first. Separating columns typically comprise a C8 or C18 carbon-chain bonded to a silica particle substrate.

Hydrophobic Interaction Chromatography

Hydrophobic interactions between proteins and the chromatographic matrix can be exploited to purify proteins. In hydrophobic interaction chromatography the matrix material is lightly substituted with hydrophobic groups. These groups can range from methyl, ethyl, propyl, octyl, or phenyl groups. At high salt concentrations, non-polar sidechains on the surface on proteins "interact" with the hydrophobic groups; that is, both types of groups are excluded by the polar solvent (hydrophobic effects are augmented by increased ionic strength). Thus, the sample is applied to the column in a buffer which is highly polar. The eluant is typically an aqueous buffer with decreasing salt concentrations, increasing concentrations of detergent (which disrupts hydrophobic interactions), or changes in pH.

In general, Hydrophobic Interaction Chromatography (HIC) is advantageous if the sample is sensitive to pH change or harsh solvents typically used in other types of chromatography but not high salt concentrations. Commonly, it is the amount of salt in the buffer which is varied. In 2012, Müller and Franzreb described the effects of temperature on HIC using Bovine Serum Albumin (BSA) with four different types of hydrophobic resin. The study altered temperature as to effect the binding affinity of BSA onto the matrix. It was concluded that cycling temperature from 50 degrees to 10 degrees would not be adequate to effectively wash all BSA from the matrix but could be very effective if the column would only be used a few times. Using temperature to effect change allows labs to cut costs on buying salt and saves money.

If high salt concentrations along with temperature fluctuations want to be avoided you can use a more hydrophobic to compete with your sample to elute it. This so-called salt independent method of HIC showed a direct isolation of Human Immunoglobulin G (IgG) from serum with satisfactory yield and used Beta-cyclodextrin as a competitor to displace IgG from the matrix. This largely opens up the possibility of using HIC with samples which are salt sensitive as we know high salt concentrations precipitate proteins.

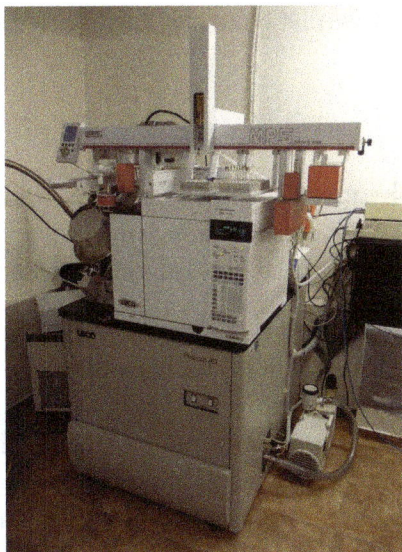

Two-dimensional chromatograph GCxGC-TOFMS at Chemical Faculty of GUT Gdańsk, Poland, 2016

Two-dimensional Chromatography

In some cases, the chemistry within a given column can be insufficient to separate some analytes. It is possible to direct a series of unresolved peaks onto a second column with different physico-chemical (Chemical classification) properties. Since the mechanism of retention on this new solid support is different from the first dimensional separation, it can be possible to separate compounds that are indistinguishable by one-dimensional chromatography. The sample is spotted at one corner of a square plate,developed, air-dried, then rotated by 90° and usually redeveloped in a second solvent system.

Simulated Moving-bed Chromatography

The simulated moving bed (SMB) technique is a variant of high performance liquid chromatography; it is used to separate particles and/or chemical compounds that would be difficult or impossible to resolve otherwise. This increased separation is brought about by a valve-and-column arrangement that is used to lengthen the stationary phase indefinitely. In the moving bed technique of preparative chromatography the feed entry and the analyte recovery are simultaneous and continuous, but because of practical difficulties with a continuously moving bed, simulated moving bed technique was proposed. In the simulated moving bed technique instead of moving the bed, the sample inlet and the analyte exit positions are moved continuously, giving the impression of a moving bed. True moving bed chromatography (TMBC) is only a theoretical concept. Its simulation, SMBC is achieved by the use of a multiplicity of columns in series and a complex valve arrangement, which provides for sample and solvent feed, and also analyte and waste takeoff at appropriate locations of any column, whereby it allows switching at regular intervals the sample entry in one direction, the solvent entry in the opposite direction, whilst changing the analyte and waste takeoff positions appropriately as well.

Pyrolysis Gas Chromatography

Pyrolysis gas chromatography mass spectrometry is a method of chemical analysis in which the sample is heated to decomposition to produce smaller molecules that are separated by gas chromatography and detected using mass spectrometry.

Pyrolysis is the thermal decomposition of materials in an inert atmosphere or a vacuum. The sample is put into direct contact with a platinum wire, or placed in a quartz sample tube, and rapidly heated to 600–1000 °C. Depending on the application even higher temperatures are used. Three different heating techniques are used in actual pyrolyzers: Isothermal furnace, inductive heating (Curie Point filament), and resistive heating using platinum filaments. Large molecules cleave at their weakest points and produce smaller, more volatile fragments. These fragments can be separated by gas chromatography. Pyrolysis GC chromatograms are typically complex because a wide range of different decomposition products is formed. The data can either be used as fingerprint to prove material identity or the GC/MS data is used to identify individual fragments to obtain structural information. To increase the volatility of polar fragments, various methylating reagents can be added to a sample before pyrolysis.

Besides the usage of dedicated pyrolyzers, pyrolysis GC of solid and liquid samples can be performed directly inside Programmable Temperature Vaporizer (PTV) injectors that provide quick heating (up to 30 °C/s) and high maximum temperatures of 600–650 °C. This is sufficient for some pyrolysis applications. The main advantage is that no dedicated instrument has to be purchased and pyrolysis can be performed as part of routine GC analysis. In this case quartz GC inlet liners have to be used. Quantitative data can be acquired, and good results of derivatization inside the PTV injector are published as well.

Fast protein Liquid Chromatography

Fast protein liquid chromatography (FPLC), is a form of liquid chromatography that is often used to analyze or purify mixtures of proteins. As in other forms of chromatography, separation is possible because the different components of a mixture have different affinities for two materials, a moving fluid (the "mobile phase") and a porous solid (the stationary phase). In FPLC the mobile phase is an aqueous solution, or "buffer". The buffer flow rate is controlled by a positive-displacement pump and is normally kept constant, while the composition of the buffer can be varied by drawing fluids in different proportions from two or more external reservoirs. The stationary phase is a resin composed of beads, usually of cross-linked agarose, packed into a cylindrical glass or plastic column. FPLC resins are available in a wide range of bead sizes and surface ligands depending on the application.

Countercurrent Chromatography

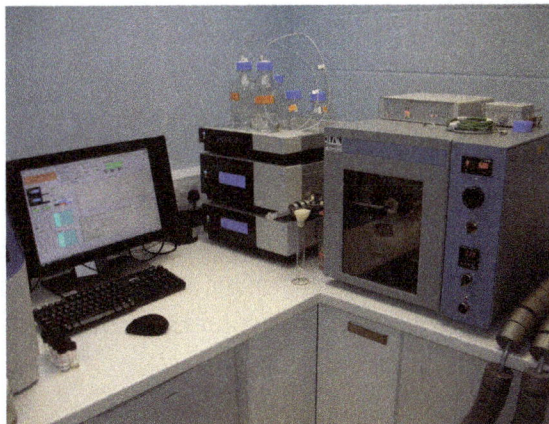

An example of a HPCCC system

Countercurrent chromatography (CCC) is a type of liquid-liquid chromatography, where both the stationary and mobile phases are liquids. The operating principle of CCC equipment requires a column consisting of an open tube coiled around a bobbin. The bobbin is rotated in a double-axis gyratory motion (a cardioid), which causes a variable gravity (G) field to act on the column during each rotation. This motion causes the column to see one partitioning step per revolution and components of the sample separate in the column due to their partitioning coefficient between the two immiscible liquid phases used. There are many types of CCC available today. These include HSCCC (High Speed CCC) and HPCCC (High Performance CCC). HPCCC is the latest and best performing version of the instrumentation available currently.

Chiral Chromatography

Chiral chromatography involves the separation of stereoisomers. In the case of enantiomers, these have no chemical or physical differences apart from being three-dimensional mirror images. Conventional chromatography or other separation processes are incapable of separating them. To enable chiral separations to take place, either the mobile phase or the stationary phase must themselves be made chiral, giving differing affinities between the analytes. Chiral chromatography HPLC columns (with a chiral stationary phase) in both normal and reversed phase are commercially available.

Protein Isolation and Characterization - II

The most convenient and useful methods for fractionating proteins make use of column chromatography. This method involves passing the protein through a column filled with resins of unique characteristics features. This takes advantage of differences in protein size, charge, binding affinity and other properties. A solid, porous material (resin) is considered as a stationary phase and the protein from a mixture is eluted through buffered solution (mobile phase) stepwise. The different types of chromatography are discussed as follows.

Ion Exchange Chromatography

This method takes the advantage of protein's surface charge. Depending upon surface residue of protein and pH of the buffered solution, protein may acquire positive or negative charges. If a protein has an isoelectric point at 7.2, it will achieve net positive at pH 6. Therefore it obviously binds with column of beads containing negatively charged groups (for example carboxylate) whereas negatively charged proteins will not bind anyway. Therefore such negatively charged protein will elute earlier. The positively

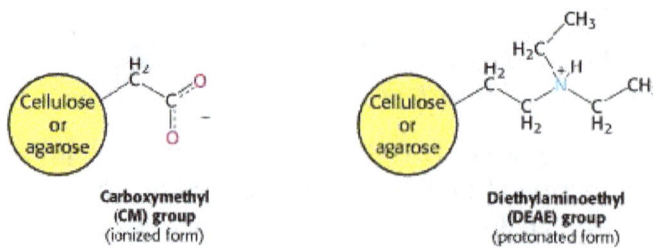

Schematic diagram of cationic and anionic exchange resin

charged protein already bound to such a column can then be eluted by increasing the concentration of sodium chloride or any another salt in the eluting buffer. Sodium ions compete with positively charged groups on the protein for binding to the column and free the protein. Carboxymethyl-cellulose (CM-cellulose) is well known cationic exchange resin. Negatively charged proteins (anionic proteins) can be separated by chromatography on positively charged diethylaminoethyl-cellulose (DEAE-cellulose)

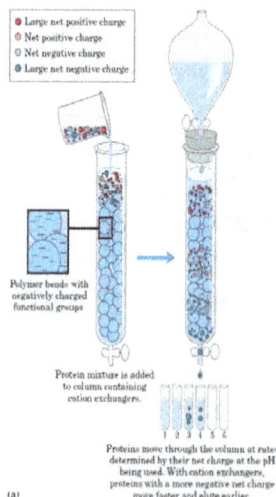

Schematic diagram of Ion exchange chromatography

columns. When a mixture contains a number of proteins, depending upon the net charge of protein in buffered solution of definite pH, nature of resin will select. After successful binding with matrix, the column is washed and the bound proteins are eluted depending on their tightness of binding, by subjecting them to either increasing concentrations of salt (sodium chloride) or changes in pH. Proteins with low charge will elute first.

Gel-Filtration Chromatography or Size exclusion chromatography

In this chromatography proteins are purified on the basis of its size. The size of the protein depends upon the number of amino acids it contains. The protein sample is applied to the column containing of an insoluble beads of engineered pores. Usually those beads are highly hydrated, cross linked polymer such as dextran or agarose (which are carbohydrates) or polyacrylamide. Most commonly available beads are Sephadex, Sepharose, and Bio-gel which are typically 100 µm in diameter. Small protein molecules can enter freely to the pores of those beads, but large ones cannot. Hence, small protein molecules are in the aqueous solution bind with the beads, whereas large protein molecules cannot bind or loosely bind with the beads. As a result of this large molecules take a short path and elute more rapidly through the column. Hence, in the elution profile, the larger protein molecules would be eluted first and the smallest ones will be last to exit.

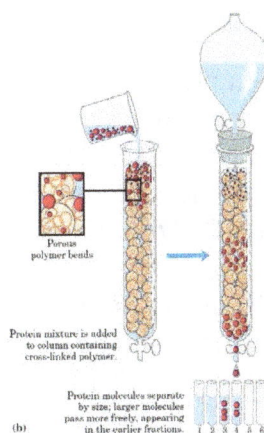

Schematic diagram of Size exclusion chromatography

Affinity Chromatography

As the name says, affinity chromatography uses the principle of binding affinity of a protein to specific chemical groups. A particular protein with affinity for a particular chemical group will bind to the beads in the column, and as a result, its migration will be hindered. For example, the plant protein Concanavalin A can be purified by passing a crude extract through a column of beads, covalently attached glucose residues. Concanavalin A binds to such a column because it has affinity for glucose, whereas most other proteins do not. The bound Concanavalin A can then be released from the column by adding a concentrated solution of glucose. The glucose in solution displaces the column-attached glucose residues from binding sites on Concanavalin A. This method of purification is used mostly in the penultimate stages where the protein is relatively pure, and needs further purification. Affinity chromatography is widely used for isolating proteins that regulate gene expression by binding to specific DNA sequences, transcription factors. Affinity chromatography is most effective when the interaction of the protein and the molecule that is used as the inducement.

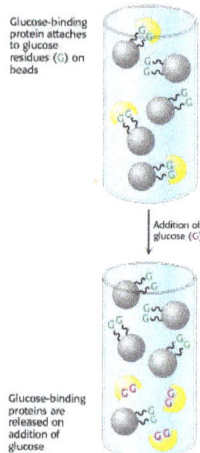

Affinity Chromatography

An extension of affinity chromatography is used for purification of recombinant proteins. The gene encoding a protein of interest is cloned into an expression vector, often tagged with a Glutathione Transferase; GST or His. This tagged protein is then introduced into the producer cell; for example bacterial (e.g. BL21), yeast (e.g. S. cerevisiae), insect (e.g. sf 9) or mammalian (e.g. CHO) cell system in order to express the protein as a fusion protein. The specific tag on that particular protein serves as a drag down. Here stationary phase is Ni-NTA (Ni-nitrilotriacetic acid) which binds selectively to His or Glutathione. Pure protein from column can be eluted by running either imidazole or pure His solution (for Ni-NTA columns) in different concentrations or Gln solution (for GST columns).

High-Pressure Liquid Chromatography (HPLC)

A current strategy in chromatographic methods is High Performance Liquid Chromatography or HPLC. The resolving power of all of the column techniques can be improved substantially through the use of a technique called high-pressure liquid chromatography (HPLC), which is an enhanced version of the column techniques already discussed. The column materials themselves are much more finely divided and, therefore there are more interaction sites with samples and thus greater resolving power. Because the column is made of finer material, pressure must be applied to the column to obtain adequate flow rates. The net result is high resolution as well as rapid separation.

Beer–Lambert Law

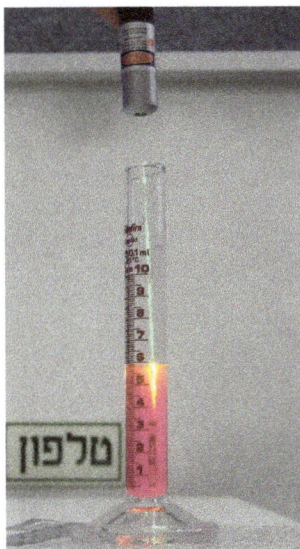

A demonstration of the Beer–Lambert law: green laser light in a solution of Rhodamine 6B. The beam radiant power becomes weaker as it passes through solution

The Beer–Lambert law, also known as Beer's law, the Lambert–Beer law, or the Beer–Lambert–Bouguer law relates the attenuation of light to the properties of the material through which the light is traveling. The law is commonly applied to chemical analysis measurements and used in understanding attenuation in physical optics, for photons, neutrons or rarefied gases. In mathematical physics, this law arises as a solution of the BGK equation.

History

The law was discovered by Pierre Bouguer before 1729. It is often attributed to Johann Heinrich Lambert, who cited Bouguer's *Essai d'optique sur la gradation de la lumière* (Claude Jombert, Paris, 1729)—and even quoted from it—in his *Photometria* in 1760. Lambert's law stated that absorbance of a material sample is directly proportional to its thickness (path length). Much later, August Beer discovered another attenuation relation in 1852. Beer's law stated that absorbance is proportional to the concentrations of the attenuating species in the material sample. The modern derivation of the Beer–Lambert law combines the two laws and correlates the absorbance to both the concentrations of the attenuating species as well as the thickness of the material sample.

Beer–Lambert Law

By definition, the transmittance of material sample is related to its optical depth τ and to its absorbance A as

$$T = \frac{\Phi_e^t}{\Phi_e^i} = e^{-\tau} = 10^{-A},$$

where

- Φ_e^t is the radiant flux *transmitted* by that material sample;

- Φ_e^i is the radiant flux received by that material sample.

The Beer–Lambert law states that, for N attenuating species in the material sample,

$$T = e^{-\sum_{i=1}^{N} \sigma_i \int_0^{\ell} n_i(z)dz} = 10^{-\sum_{i=1}^{N} \varepsilon_i \int_0^{\ell} c_i(z)dz},$$

or equivalently that

$$\tau = \sum_{i=1}^{N} \tau_i = \sum_{i=1}^{N} \sigma_i \int_0^{\ell} n_i(z)dz,$$

$$A = \sum_{i=1}^{N} A_i = \sum_{i=1}^{N} \varepsilon_i \int_0^{\ell} c_i(z)dz,$$

where

- σ_i is the attenuation cross section of the attenuating species i in the material sample;

- n_i is the number density of the attenuating species i in the material sample;

- ε_i is the molar attenuation coefficient or absorptivity of the attenuating species i in the material sample;

- c_i is the amount concentration of the attenuating species i in the material sample;

- ℓ is the path length of the beam of light through the material sample.

Attenuation cross section and molar attenuation coefficient are related by

$$\varepsilon_i = \frac{N_A}{\ln 10} \sigma_i,$$

and number density and amount concentration by

$$c_i = \frac{n_i}{N_A},$$

where N_A is the Avogadro constant.

In case of *uniform* attenuation, these relations become

$$T = e^{-\sum_{i=1}^{N} \sigma_i n_i \ell} = 10^{-\sum_{i=1}^{N} \varepsilon_i c_i \ell},$$

or equivalently

$$\tau = \sum_{i=1}^{N} \sigma_i n_i \ell,$$

$$A = \sum_{i=1}^{N} \varepsilon_i c_i \ell.$$

Cases of *non-uniform* attenuation occur in atmospheric science applications and radiation shielding theory for instance.

The law tends to break down at very high concentrations, especially if the material is highly scattering. If the radiation is especially intense, nonlinear optical processes can also cause variances. The main reason, however, is the following. At high concentrations, the molecules are closer to each other and begin to interact with each other. This interaction will change several properties of the molecule, and thus will change the attenuation.

Expression with Attenuation Coefficient

The Beer–Lambert law can be expressed in terms of attenuation coefficient, but in this case is better called Lambert's law since amount concentration, from Beer's law, is hidden inside the attenuation coefficient. The (Napierian) attenuation coefficient μ and the decadic attenuation coefficient $\mu_{10} = \mu / \ln 10$ of a material sample are related to its number densities and amount concentrations as

$$\mu(z) = \sum_{i=1}^{N} \mu_i(z) = \sum_{i=1}^{N} \sigma_i n_i(z),$$

$$\mu_{10}(z) = \sum_{i=1}^{N} \mu_{10,i}(z) = \sum_{i=1}^{N} \varepsilon_i c_i(z)$$

respectively, by definition of attenuation cross section and molar attenuation coefficient. Then the Beer–Lambert law becomes

$$T = e^{-\int_0^\ell \mu(z)\,\mathrm{d}z} = 10^{-\int_0^\ell \mu_{10}(z)\,\mathrm{d}z},$$

and

$$\tau = \int_0^\ell \mu(z)\mathrm{d}z,$$

$$A = \int_0^\ell \mu_{10}(z)\mathrm{d}z.$$

In case of *uniform* attenuation, these relations become

$$T = e^{-\mu\ell} = 10^{-\mu_{10}\ell},$$

or equivalently

$$\tau = \mu\ell,$$

$$A = \mu_{10}\ell.$$

Derivation

Assume that a beam of light enters a material sample. Define z as an axis parallel to the direction of the beam. Divide the material sample into thin slices, perpendicular to the beam of light, with thickness dz sufficiently small that one particle in a slice cannot obscure another particle in the same slice when viewed along the z direction. The radiant flux of the light that emerges from a slice is reduced, compared to that of the light that entered, by $d\Phi_e(z) = -\mu(z)\Phi_e(z)\,dz$, where μ is the (Napierian) attenuation coefficient, which yields the following first-order linear ODE:

$$\frac{\mathrm{d}\Phi_e}{\mathrm{d}z}(z) = -\mu(z)\Phi_e(z).$$

The attenuation is caused by the photons that did not make it to the other side of the slice because of scattering or absorption. The solution to this differential equation is obtained by multiplying the integrating factor

$$e^{\int_0^z \mu(z')\mathrm{d}z'}$$

throughout to obtain

$$\frac{\mathrm{d}\Phi_e}{\mathrm{d}z}(z)e^{\int_0^z \mu(z')\mathrm{d}z'} + \mu(z)\Phi_e(z)e^{\int_0^z \mu(z')\mathrm{d}z'} = 0,$$

which simplifies due to the product rule (applied backwards) to

$$\frac{\mathrm{d}}{\mathrm{d}z}\left(\Phi_e(z)e^{\int_0^z \mu(z')\mathrm{d}z'}\right) = 0.$$

Integrating both sides and solving for Φ_e for a material of real thickness ℓ, with the incident radiant flux upon the slice $\Phi_e{}^i = \Phi_e(0)$ and the transmitted radiant flux $\Phi_e{}^t = \Phi_e(\ell)$ gives

$$\Phi_e^t = \Phi_e^i e^{-\int_0^\ell \mu(z)dz},$$

and finally

$$T = \frac{\quad}{\quad} = e^{\int (z)dz}$$

Since the decadic attenuation coefficient μ_{10} is related to the (Napierian) attenuation coefficient by $\mu_{10} = \mu/\ln 10$, one also have

$$T = e^{-\int_0^\ell \ln 10\,\mu_{10}(z)dz} = \left(e^{-\int_0^\ell \mu_{10}(z)dz}\right)^{\ln 10} = 10^{-\int_0^\ell \mu_{10}(z)dz}.$$

To describe the attenuation coefficient in a way independent of the number densities n_i of the N attenuating species of the material sample, one introduces the attenuation cross section $\sigma_i = \mu_i(z)/n_i(z)$. σ_i has the dimension of an area; it expresses the likelihood of interaction between the particles of the beam and the particles of the specie i in the material sample:

$$T = e^{-\sum_{i=1}^{N} \sigma_i \int_0^\ell n_i(z)dz}.$$

One can also use the molar attenuation coefficients $\varepsilon_i = (N_A/\ln 10)\sigma_i$, where N_A is the Avogadro constant, to describe the attenuation coefficient in a way independent of the amount concentrations $c_i(z) = n_i(z)/N_A$ of the attenuating species of the material sample:

$$T = e^{-\sum_{i=1}^{N} \frac{\ln 10}{N_A}\varepsilon_i \int_0^\ell n_i(z)dz} = \left(e^{-\sum_{i=1}^{N} \varepsilon_i \int_0^\ell \frac{n_i(z)}{N_A}dz}\right)^{\ln 10} = 10^{-\sum_{i=1}^{N} \varepsilon_i \int_0^\ell c_i(z)dz}.$$

Validity

Under certain conditions Beer–Lambert law fails to maintain a linear relationship between attenuation and concentration of analyte. These deviations are classified into three categories:

1. Real–fundamental deviations due to the limitations of the law itself.

2. Chemical–deviations observed due to specific chemical species of the sample which is being analyzed.

3. Instrument—deviations which occur due to how the attenuation measurements are made.

There are at least six conditions that need to be fulfilled in order for Beer–Lambert law to be valid. These are:

1. The attenuators must act independently of each other.

2. The attenuating medium must be homogeneous in the interaction volume.

3. The attenuating medium must not scatter the radiation—no turbidity—unless this is accounted for as in DOAS.

4. The incident radiation must consist of parallel rays, each traversing the same length in the absorbing medium.

5. The incident radiation should preferably be monochromatic, or have at least a width that is narrower than that of the attenuating transition. Otherwise a spectrometer as detector for the power is needed instead of a photodiode which has not a selective wavelength dependence.

6. The incident flux must not influence the atoms or molecules; it should only act as a non-invasive probe of the species under study. In particular, this implies that the light should not cause optical saturation or optical pumping, since such effects will deplete the lower level and possibly give rise to stimulated emission.

If any of these conditions are not fulfilled, there will be deviations from Beer–Lambert law.

The Beer–Lambert law is not compatible with Maxwell's equations. Being strict, the law does not describe the transmittance through a medium, but the propagation within that medium. It can be made compatible with Maxwell's equations if the transmittance of a sample with solute is ratioed against the transmittance of the pure solvent which explains why it works so well in spectrophotometry. As this is not possible for pure media, the uncritical employment of the Beer–Lambert law can easily generate errors of the order of 100% or more. In such cases it is necessary to apply the Transfer-matrix method.

Chemical Analysis by Spectrophotometry

Beer–Lambert law can be applied to the analysis of a mixture by spectrophotometry, without the need for extensive pre-processing of the sample. An example is the determination of bilirubin in blood plasma samples. The spectrum of pure bilirubin is known, so the molar attenuation coefficient ε is known. Measurements of decadic attenuation coefficient μ_{10} are made at one wavelength λ that is nearly unique for bilirubin and at a second wavelength in order to correct for possible interferences. The amount concentration c is then given by

$$c = \frac{\mu_{10}(\lambda)}{\varepsilon(\lambda)}.$$

For a more complicated example, consider a mixture in solution containing two species at amount concentrations c_1 and c_2. The decadic attenuation coefficient at any wavelength λ is, given by

$$\mu_{10}(\lambda) = \varepsilon_1(\lambda)c_1 + \varepsilon_2(\lambda)c_2.$$

Therefore, measurements at two wavelengths yields two equations in two unknowns and will suffice to determine the amount concentrations c_1 and c_2 as long as the molar attenuation coefficient of the two components, ε_1 and ε_2 are known at both wavelengths. This two system equation can be solved using Cramer's rule. In practice it is better to use linear least squares to determine the two amount concentrations from measurements made at more than two wavelengths. Mixtures containing more than two components can be analyzed in the same way, using a minimum of N wavelengths for a mixture containing N components.

The law is used widely in infra-red spectroscopy and near-infrared spectroscopy for analysis of polymer degradation and oxidation (also in biological tissue) as well as to measure the concentration of various compounds in different food samples. The carbonyl group attenuation at about 6 micrometres can be detected quite easily, and degree of oxidation of the polymer calculated.

Beer–Lambert law in the Atmosphere

This law is also applied to describe the attenuation of solar or stellar radiation as it travels through the atmosphere. In this case, there is scattering of radiation as well as absorption. The optical depth for a slant path is $\tau' = m\tau$, where τ refers to a vertical path, m is called the relative airmass, and for a plane-parallel atmosphere it is determined as $m = \sec\theta$ where θ is the zenith angle corresponding to the given path. The Beer–Lambert law for the atmosphere is usually written

$$T = e^{-m(\tau_a + \tau_g + \tau_{RS} + \tau_{NO_2} + \tau_w + \tau_{O_3} + \tau_r + \dots)},$$

where each τ_x is the optical depth whose subscript identifies the source of the absorption or scattering it describes:

- a refers to aerosols (that absorb and scatter);

- g are uniformly mixed gases (mainly carbon dioxide (CO_2) and molecular oxygen (O_2) which only absorb);

- NO_2 is nitrogen dioxide, mainly due to urban pollution (absorption only);

- RS are effects due to Raman scattering in the atmosphere;

- w is water vapour absorption;

- O_3 is ozone (absorption only);

- r is Rayleigh scattering from molecular oxygen (O_2) and nitrogen (N_2) (responsible for the blue color of the sky);

- the selection of the attenuators which have to be considered depends on the wavelength range and can include various other compounds. This can include tetraoxygen, HONO, formaldehyde, glyoxal, a series of halogen radicals and others.

m is the *optical mass* or *airmass factor*, a term approximately equal (for small and moderate values of θ) to $1/\cos \theta$, where θ is the observed object's zenith angle (the angle measured from the direction perpendicular to the Earth's surface at the observation site). This equation can be used to retrieve τ_a, the aerosol optical thickness, which is necessary for the correction of satellite images and also important in accounting for the role of aerosols in climate.

Protein isolation and characterization - III

Prior to any characterization procedure of the protein, it is necessary to determine its concentration. Several methods are used for this purpose depending on the amount of sample, accuracy and presence of interfering agents.

The various methods and their specifications are discussed below.

Spectrophotometric Determination of Protein Concentration

According to Beer-Lambert's Law, absorbance (A) has a linear relationship with molar concentration (c)

$$A = \varepsilon \times c \times l$$

where ε is the molar absorption coefficient ($M^{-1}cm^{-1}$) and l is the cell path length (cm). It is important to have an accurate value of ε to determine c from A.

Absorbance Assay

Almost every protein shows a characteristic absorption maximum of 280 nm, primarily due to amino acids residues containing aromatic ring (like Trp, Tyr and a lesser extent, Phe). Whereas the peptide bonds absorb at around 205 nm. This unique absorbance property of proteins is widely used to estimate the quantity of proteins.

Those methods are quite accurate with the ranges of protein from 20µg to 3mg of for absorbance at 280 nm, and with 1 to 100µg for absorbance at 205 nm. Absorbance of a known concentration of a protein solution was determined spectrophotometrically either at 280 nm or at 205 nm, knowing the ε value of the same protein at respective wavelength, its concentration can be determined.

The assay is non-destructive, in most cases protein is not consumed and can be recovered after analysis.

Bradford Protein Assay

This assay is applied for its simplicity, scalability and sensitivity. The basis of this assay is absorbance maximum of an acidic solution of Coomassie Brilliant Blue G- 250 shifts from 465nm to 595nm upon protein binding. Both hydrophobic and ionic interactions stabilize the anionic form of the dye, causing a visible color change. Thus from the absorbance at 595 nm, amount of protein can be determined quantitatively.

A calibration curve is made by adding known concentration of bovine serum albumin with Bradford's reagent (containing Coomassie Brilliant Blue G-250, ethanol, phosphoric acid). Finally the same Bradford's reagent is treated with unknown protein and measured the absorbance. From calibration curve, concentration can be determined.

Protein Separation and Its Characterization

An important technique for the separation of proteins is used which is based on the migration of charged proteins under an application of an electric field. Such process called electrophoresis and it is extensively used for separating proteins and other macromolecules, such as DNA and RNA. These procedures are not generally used to purify proteins because simpler alternatives are usually available (like various types of chromatography) and electrophoretic methods often affect the structure and hence, the biological function of proteins. The velocity of migration (v) of a Protein in an electric field depends on the electric field strength (E), the net charge on the protein (z), and the frictional coefficient (f)

$$v = \frac{Ez}{f}$$

And the frictional coefficient (f) depends on both the mass (m) and shape of the migrating molecule and the viscosity (η) of the medium. For a sphere of radius r,

$$f = 6\pi\eta r$$

Electrophoresis is useful as an analytical method. The advantage of electrophoresis technique is that proteins can be visualized as well as separated. This allows estimating the number of protein in a mixture as well as its purity. Electrophoresis also permits determination of approximate molecular weight and isoelectric point of a protein. Electrophoretic separations are usually carried out in gels (often solid supports such as paper). The gel serves as a molecular sieve that enhances separation protein molecules that are enough small as compared with the pores in the gel, readily move through the gel, whereas protein molecules of much larger than the pores are hindered and therefore their mobility becomes hindered.

Sodium Dodecyl Sulfate - polyacrylamide Gel Electrophoresis (SDS-PAGE)

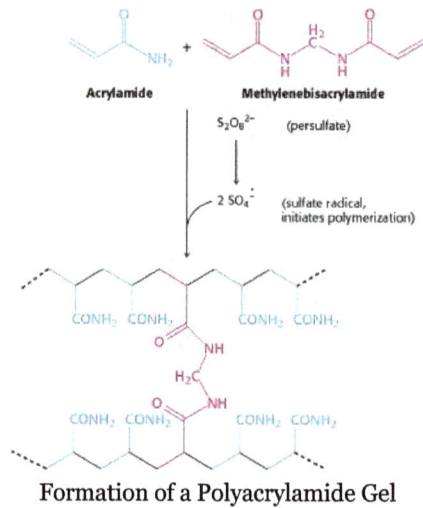

Formation of a Polyacrylamide Gel

In this type of gel electrophoresis, an anionic detergent, SDS is used, which binds to almost every protein, for every two amino acid residues. This affects unfolding of protein by destruction of hydrophobic interaction and hydrogen bonding and protein acquire a net negative charge. β-mercaptoethanol (BME) or dithiothreitol (DTT) also is added to reduce disulfide bonds. Gel electrophoresis in presence of SDS separates proteins almost on the basis of their molecular weight. That is protein having smaller molecular weight moves more rapidly.

Electrophoresis is performed in a thin, vertical slab of polyacrylamide (a polymeric gel made by treating acrylamide and methylenebis-acrylamide). On application of electric field, protein flows from top to bottom of the gel.

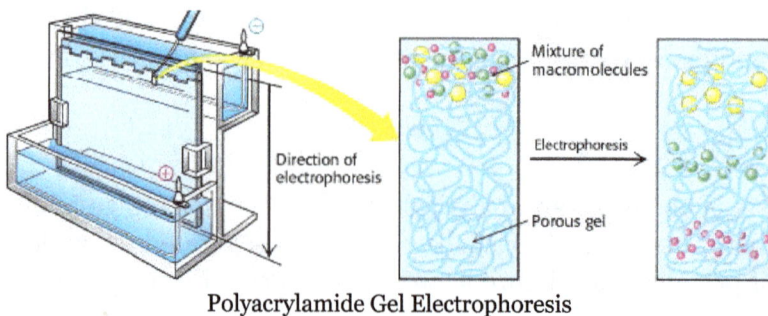

Polyacrylamide Gel Electrophoresis

After separation in gel electrophoresis, separated proteins are visualized after staining the gel. Different strainers such as Coomassie blue stain or more sensitive silver staining are used for this purpose. The Coomassie blue staining is relatively less sensitive than silver staining, but Coomassie blue staining is applied for regular staining. Coomassie blue selectively binds with protein in gel. When the gel is distained by methanol-acetic acid solution, bands of separated protein are visualized. In the case of silver staining, the gel is soaked with soluble silver ions (a solution of silver nitrate) and developed by treatment with formaldehyde, which reduces silver ions to form an insoluble brown precipitate of metallic silver. This reduction is facilitated by protein.

Isoelectric Focusing

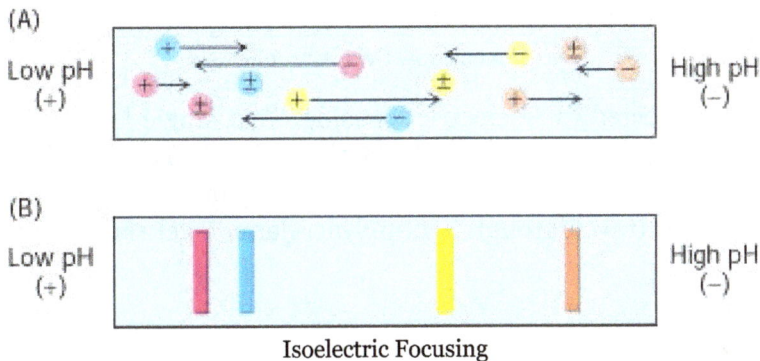

Isoelectric Focusing

Proteins can also be separated electrophoretically depending of their relative contents of acidic and basic residues. The isoelectric point (pI) of a protein is the pH at which its net charge is zero. At isoelectric point, its electrophoretic mobility is zero. For example, if a mixture of human serum albumin (pI 4.9, acidic protein) and human gamma crystallin (pI 7.2, basic protein) is subjected to isoelectric focusing, in a pH gradient gel in the absence of SDS, each protein will move until it reaches a position in the gel at which the pH is equal to the pI of the protein. Proteins with different isoelectric points are thus distributed differently throughout the gel. This method of separating proteins according to their isoelectric point is called isoelectric focusing.

A pH gradient gel is made by mixing low molecular weight organic acids and bases (often called ampholytes) to distribute them in an electric field generated across the gel. Isoelectric focusing can readily resolve proteins that differ in pI by as little as 0.01.

Two-Dimensional Electrophoresis

A combination of SDS-PAGE and isoelectric focusing is called two dimensional electrophoresis (2D electrophoresis). Proteins isolated from cells in different physiological conditions can be subjected to two-dimensional electrophoresis for its analysis. This gives a better resolution in terms of protein separation.

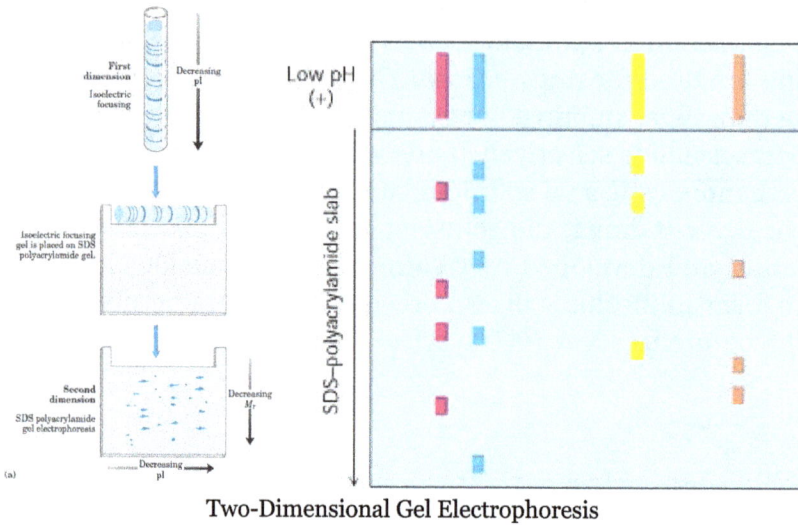

First
dimension   Decreasing
                        pI   Low pH
Isoelectric   (+)
focusing

Isoelectric focusing
gel is placed on SDS
polyacrylamide gel.

Second
dimension   Decreasing
                              M_r
SDS polyacrylamide
gel electrophoresis

(a)
        Decreasing
           pI

Two-Dimensional Gel Electrophoresis

A sample is first subjected to isoelectric focusing. That single lane gel is then placed horizontally on the top of an SDS-polyacrylamide gel slab. The proteins are therefore separated across the polyacrylamide gel according they migrated during isoelectric focusing. When protein travels trough SDS-polyacrylamide gel slab, it separates on the basis their mass.

Western Blotting

Western Blotting or immunoblotting permits determination of the presence of a specific protein in a sample after separation on SDSPAGE, for example such as a viral protein in the blood. The term Western blotting is used after a similar term 'Southern blotting' (for detection of nucleic acids in a blot technique), which was coined by and named after E. M. Southern.

The proteins are separated on SDS-PAGE and then transferred to a membrane of nitrocellulose. The membrane is incubated with a source of non-specific Western Blotting protein to bind to any remaining places on the membrane. A primary antibody is then added to the solution which is able to bind to its specific target (antigen) protein followed by washes and further incubation in a solution of secondary antibody. The secondary antibody recognizes the primary antibody and binds at locations on the blot where the primary antibody is bound as well. Generally the secondary antibody is tagged with an enzyme or marked with fluorescence probe for detection.

Centrifugation

Centrifugation is a process which involves the application of the centrifugal force for the sedimentation of heterogeneous mixtures with a centrifuge, and is used in industrial and laboratory settings. This process is used to separate two miscible substances, but also to analyze the hydrodynamic properties of macromolecules. More-dense components of the mixture migrate away from the axis of the centrifuge, while less-dense components of the mixture migrate towards the axis. Chemists and biologists may increase the effective gravitational force on a test tube so as to more rapidly and completely cause the precipitate (pellet) to gather on the bottom of the tube. The remaining solution (supernatant) may be discarded with a pipette.

There is a correlation between the size and density of a particle and the rate that the particle separates from a heterogeneous mixture, when the only force applied is that of gravity. The larger the size and the larger the density of the particles, the faster they separate from the mixture. By applying a larger effective gravitational force to the mixture, like a centrifuge does, the separation of the particles is accelerated. This is ideal in industrial and lab settings because particles that would naturally separate over a long period of time can be separated in much less time.

The rate of centrifugation is specified by the angular velocity usually expressed as revolutions per minute (RPM), or acceleration expressed as g. The conversion factor between RPM and g depends on the radius of the centrifuge rotor. The particles' settling velocity in centrifugation is a function of their size and shape, centrifugal acceleration, the volume fraction of solids present, the density difference between the particle and the liquid, and the viscosity. The most common application is the separation of solid from highly concentrated suspensions, which is used in the treatment of sewage sludges for dewatering where less consistent sediment is produced.

In the chemical and food industries, special centrifuges can process a continuous stream of particle-laden liquid.

Centrifugation is the most common method used for uranium enrichment, relying on the slight mass difference between atoms of U238 and U235 in uranium hexafluoride gas.

Mathematical Formula

The general formula for calculating the revolutions per minute (RPM) of a centrifuge is

$$RPM = \sqrt{\frac{g}{r}},$$

where g represents the respective force of the centrifuge and r the radius from the center of the rotor to a point in the sample. However, depending on the centrifuge model used, the respective angle of the rotor and the radius may vary, thus the formula gets modified. For example, the Sorvall #SS-34 rotor has a maximum radius of 10.8 cm, so the formula becomes $RPM = 299\sqrt{\frac{g}{r}}$, which can further simplify to $RPM = 91\sqrt{g}$.

Centrifugation in Biological Research

Microcentrifuges

Microcentrifuges are used to process small volumes of biological molecules, cells, or nuclei. Microcentrifuge tubes generally hold 0.5 - 2.0 mL of liquid, and are spun at maximum angular speeds of 12,000–13,000 rpm. Microcentrifuges are small enough to fit on a table-top and have rotors that can quickly change speeds. They may or may not have a refrigeration function.

High-speed Centrifuges

High-speed or superspeed centrifuges can handle larger sample volumes, from a few tens of millilitres to several litres. Additionally, larger centrifuges can also reach higher angular velocities (around 30,000 rpm). The rotors may come with different adapters to hold various sizes of test tubes, bottles, or microtiter plates.

Fractionation Process

General method of fractionation: Cell sample is stored in a suspension which is:

1. Buffered - neutral pH, preventing damage to the structure of proteins including enzymes (which could affect ionic bonds)

2. Isotonic (of equal water potential) - this prevents water gain or loss by the organelles

3. Cool - reducing the overall activity of enzyme released later in the procedure

- Cells are homogenised in a blender and filtered to remove debris

- The homogenised sample is placed in an ultracentrifuge and spun in low speed - nuclei settle out, forming a pellet

- The supernatant (suspension containing remaining organelles) is spun at a higher speed - chloroplasts settle out

- The supernatant is spun at a higher speed still - mitochondria and lysosomes settle out

- The supernatant is spun at an even higher speed - ribosomes, membranes settle out

The ribosomes, membranes and Golgi complexes can be separated by another technique called density gradient centrifugation.

Ultracentrifugations

Ultracentrifugation makes use of high centrifugal force for studying properties of biological particles. Compared to microcentrifuges or high-speed centrifuges, ultracentrifuges can isolate much smaller particles, including ribosomes, proteins, and viruses. Ultracentrifuges can also be used in the study of membrane fractionation. This occurs because ultracentrifuges can reach maximum angular velocities in excess of 70,000 rpm. Additionally, while microcentrifuges and supercentrifuges separate particles in batches (limited volumes of samples must be handled manually in test tubes or bottles), ultracentrifuges can separate molecules in batch or continuous flow systems.

In addition to purification, analytical ultracentrifugation (AUC) can be used for determination of the properties of macromolecules such as shape, mass, composition, and conformation. Samples are centrifuged with a high-density solution such as sucrose, caesium chloride, or iodixanol. The high-density solution may be at a uniform concentration throughout the test tube ("cushion") or a varying concentration ("gradient"). Molecular properties can be modeled through sedimentation velocity analysis or sedimentation equilibrium analysis. During the run, the particle or molecules will migrate through the test tube at different speeds depending on their physical properties and the properties of the solution, and eventually form a pellet at the bottom of the tube, or bands at various heights.

Density Gradient Centrifugation

Density gradient centrifugation Is considered one of the more efficient methods of separating suspended particles. Density gradient centrifugation can be used both as a separation technique and as a method of measuring the densities of particles or molecules in a mixture. A tube, after being centrifuged by this method, has particles in order of density based on height. The object or particle of interest will reside in the position within the tube corresponding to its density.

Linderstorm-Lang, in 1937, discovered that density gradient tubes could be used for density measurements. He discovered this when working with potato yellow-dwarf virus.

This method was also used in Meselson and Stahl's famous experiment in which they proved that DNA replication is semi-conservative by using different isotopes of nitrogen. They used density gradient centrifugation to determine which isotope or isotopes of nitrogen were present in the DNA after cycles of replication.

Nevertheless, some non-ideal sedimentations are still possible when using this method. The first potential issue is the unwanted aggregation of particles, but this can occur in any centrifugation. The second possibility occurs when droplets of solution that contain particles sediment. This is more likely to occur when working with a solution that has a layer of suspension floating on a dense liquid, which in fact have little to no density gradient.

Differential Centrifugation

Differential Centrifugation is a type of centrifugation in which one selectively spins down components of a mixture by a series of increasing centrifugation forces. This method is commonly used to separate organelles and membranes found in cells. Organelles generally differ from each other in density in size, making the use of differential centrifugation, and centrifugation in general, possible. The organelles can then be identified by testing for indicators that are unique to the specific organelles.

Other Applications

- Separating chalk powder from water

- Removing fat from milk to produce skimmed milk

- Separating particles from an air-flow using cyclonic separation

- The clarification and stabilization of wine

- Separation of urine components and blood components in forensic and research laboratories

- Aids in separation of proteins using purification techniques such as salting out, e.g. ammonium sulfate precipitation.

History

By 1923 Theodor Svedberg and his student H. Rinde had successfully analyzed large-grained sols in terms of their gravitational sedimentation. Sols consist of a substance evenly distributed in another substance, also known as a colloid. However, smaller grained sols, such as those containing gold, could not be analyzed. To investigate this problem Svedberg developed an analytical centrifuge, equipped with a photographic absorption system, which would exert a much greater centrifugal effect. In addition, he developed the theory necessary to measure molecular weight. During this time, Svedberg's attention shifted from gold to proteins.

By 1900, it had been generally accepted that proteins were composed of amino acids; however, whether proteins were colloids or macromolecules was still under debate. One protein being investigated at the time was hemoglobin. It was determined to have 712 carbon, 1,130 hydrogen, 243 oxygen, two sulfur atoms, and at least one iron atom. This gave hemoglobin a resulting weight of approximately 16,000 dalton (Da) but it was uncertain whether this value was a multiple of one or four (dependent upon the number of iron atoms present).

Through a series of experiments utilizing the sedimentation equilibrium technique, two important observations were made: hemoglobin has a molecular weight of 68,000 Da, suggesting that there are four iron atoms present rather than one, and that no matter where the hemoglobin was isolated from, it had exactly the same molecular weight. How something of such a large molecular mass could be consistently found, regardless of where it was sampled from in the body, was unprecedented and favored the idea the proteins are macromolecules rather than colloids. In order to investigate this phenomenon, a centrifuge with even higher speeds was needed, and thus the ultracentrifuge was created to apply the theory of sedimentation-diffusion. The same molecular mass was determined, and the presence of a spreading boundary suggested that it was a single compact particle. Further application of centrifugation showed that under different conditions the large homogeneous particles could be broken down into discrete subunits. The development of centrifugation was a great advance in experimental protein science.

References

- Simoni, D. S., Hill, R. L., and Vaughan, M. (2002). The structure and function of hemoglobin: Gilbery Smithson Adair and the Adair equations. The Journal of Biological Chemistry. 277(31): e1-e2

- Alan D. MacNaught; Andrew R. Wilkinson, eds. (1997). Compendium of Chemical Terminology: IUPAC Recommendations (the "Gold Book"). Blackwell Science. ISBN 0865426848

- De Clercq E (October 2005). "Recent highlights in the development of new antiviral drugs". Curr. Opin. Microbiol. 8 (5): 552–60. PMID 16125443. doi:10.1016/j.mib.2005.08.010

- Quignot, Nadia; Bois, Frédéric Yves (2013). "A computational model to predict rat ovarian steroid secretion from in vitro experiments with endocrine disruptors". PLoS ONE. 8 (1): e53891. PMC 3543310. PMID 23326527. doi:10.1371/journal.pone.0053891

- Cox, David L. Nelson, Michael M. (2008). Lehninger principles of biochemistry (5th ed.). New York: W.H. Freeman. ISBN 9780716771081

- Mathers, T. L.; Schoeffler, G.; McGlynn, S. P. (July 1985). "The effects of selected gases upon ethanol: hydrogen bond breaking by O and N". Canadian Journal of Chemistry. 63 (7): 1864–1869. doi:10.1139/v85-309

- Iverson, Cheryl, et al. (eds) (2007). "12.1.1 Use of Italics". AMA Manual of Style (10th ed.). Oxford, Oxfordshire: Oxford University Press. ISBN 978-0-19-517633-9

- Wang, X (2014). "Characterization of denaturation and renaturation of DNA for DNA hybridization". Environmental Health and Toxicology Environ Health Toxicol. 29. doi:10.5620/eht.2014.29.e2014007

- Sicard, François; Destainville, Nicolas; Manghi, Manoel (21 January 2015). "DNA denaturation bubbles: Free-energy landscape and nucleation/closure rates". The Journal of Chemical Physics. 142 (3): 034903. arXiv:1405.3867. doi:10.1063/1.4905668

- American Psychological Association (2010), "4.21 Use of Italics", The Publication Manual of the American Psychological Association (6th ed.), Washington, DC, USA: APA, ISBN 978-1-4338-0562-2

- "Denaturing Polyacrylamide Gel Electrophoresis of DNA & RNA". Electrophoresis. National Diagnostics. Retrieved 13 October 2016

- Richard, C., and A. J. Guttmann. "Poland–Scheraga Models and the DNA Denaturation Transition." Journal of Statistical Physics 115.3/4 (2004): 925-47

- Iverson, Cheryl, et al. (eds) (2007). "12.1.1 Use of Italics". AMA Manual of Style (10th ed.). Oxford, Oxfordshire: Oxford University Press. ISBN 978-0-19-517633-9

- "Terminology for biorelated polymers and applications (IUPAC Recommendations 2012)" (PDF). Pure and Applied Chemistry. 84 (2): 377–410. 2012. doi:10.1351/PAC-REC-10-12-04

- American Psychological Association (2010), "4.21 Use of Italics", The Publication Manual of the American Psychological Association (6th ed.), Washington, DC, USA: APA, ISBN 978-1-4338-0562-2

- Charles Tanford (1968), "Protein denaturation" (PDF), Advances in Protein Chemistry, 23: 121–282, doi:10.1016/S0065-3233(08)60401-5, PMID 4882248

Types of Proteins

Proteins are of various types with a vast arrangement of functions within an organism. Some of the types of proteins are metalloproteins, motor proteins, membrane proteins, etc. Mostly proteins belong to the category of metalloproteins. The major categories of protein are dealt with great details in this chapter.

Metalloproteins

A large number of biochemical reactions are known to be catalyzed by proteins. The reason behind their functions as catalysts is attributed to the side chains of amino acid up to some extent and majorly to their ability to incorporate various cofactors such as metal ion, clusters, and organic molecules into their active site.

About half of available proteins require metal ions at their active site to function. Proteins possessing one or more metal ions execute ample number of functions and are known as metalloproteins or metalloenzymes. The functionality of enzymes varies with the metal centre at their active site. For instance oxidation and reduction process generally involve Fe, Mn, Cu and Mo; Co plays important role in radical-based rearrangement and methyl-group transfer reaction; Zn, Fe, Mg, Mn and Ni are important for hydrolysis process and DNA processing involves Zn. Apart from this metalloenzymes are also involved in various cell function such as storage and transport of proteins, enzymes and signal transduction proteins.

Metal-coordination Sites

The usual coordination around the metal center in metalloproteins involves nitrogen, oxygen or sulfur atoms belonging to various amino acid residues of the protein. The functional groups involved in coordination, often comes from the side chain of amino acid residues. Among those imidazole substituent in histidine residues, thiolate substituents in cysteinyl residues, and carboxylate groups provided by aspartate are the most important one. Apart from these the peptide backbone participate in coordination, generally via deprotonated amides and the carbonyl oxygen centres of amide bond.

In addition to this, a large number of organic cofactors acts as ligands. The most popular are the tetradentate N4 macrocyclic ligands incorporated into the heme protein. Inorganic ligands such as sulfide and oxide are also common.

Functional Classification of Metalloprotein

Function	Protein	Metal/Metal Complexes	PDB
Electron Transfer	Cytochrome b5 Adrenodoxin Plastocyanin	heme b Fe2S2 Cu	1CYO
			1AYF
			1AG6
Light harvesting	Light harvesting complex LH-II	BChl-a	1KZU
Catalysis	Nitrile hydratase DMSO reductase Nitrogenase MoFe protein Manganese superixide dismutase	Fe Moco FeMoco Mn	2AHJ
			1DMR
			3MIN
			1VEW
Storage (uptake, binding and release)	Nitrophorin Hemocyanin Metallothioneins Lactoferrin Bacterioferritin	Heme(coordinates NO) $2Cu^{2+}$ (co-ordinates O2) Cd^{2+}, Hg^{2+}, Pb^{2+}, Tl^{+} Fe	4NP1
			1OXY
		Fe (in form of hydrated ferric phosphate)	4MT2
			1B1X
			1BFR
Translocation	Copper transporting ATPase	Cu^{+}	2AWO
Various	Lignin peroxidase Zinc finger Endonuclease III	Ca^{2+} Zn^{2+} Fe4S4	1B82
			1AAY
			2ABK

- Storage and transport metalloprotein

– Oxygen carriers

- Hemoglobin and myoglobin

The two of the earliest structurally characterized proteins hemoglobin (Hb) and myoglobin (Mb) contain iron protoporphyrin IX (heme) as a prosthetic centre. Both the proteins bind reversibly with O_2 however their biological role is different. Hb transport oxygen in blood plasma whereas, Mb accumulates O_2 in cellular tissue. Hb contains four sub-units in which the Fe(II) ion is coordinated by the planar, macrocyclic ligand protoporphyrin IX and the imidazole nitrogen atom of a histidine residue. The sixth coordination site encloses a water molecule or a dioxygen moiety. On the other side myoglobin has only one such unit and the active site is located in a hydrophobic pocket. The four subunits of hemoglobin show cooperativity effect which allows it to transfer oxygen to myoglobin.

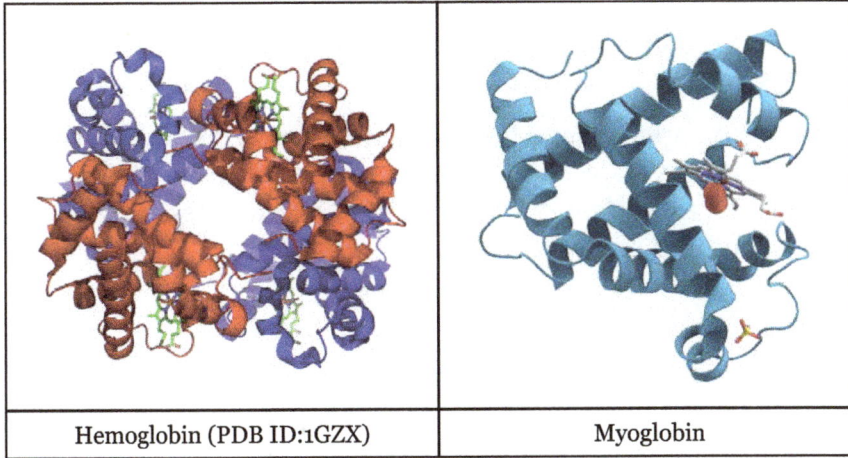

| Hemoglobin (PDB ID:1GZX) | Myoglobin |

The diamagnetic nature of both the protein is attributed to low-spin state of Fe(II). Both Hb and Mb bind O_2 in the reduced state. The iron atom is located in the plane of the porphyrin ring in oxyhemoglobin, whereas it lies above the plane of the ring in deoxyhemoglobin.

Hemerythrin and Hemocyanin

Non heme proteins like hemerythrin and hemocyanin, found only in invertebrates, are another class of oxygen carrier protein. Hemerythrin is an iron containing protein in which O_2 binds at binuclear iron center. The coordination environment around iron atoms involves carboxylate side chains of glutamate and aspartate and five histidine residues. Reduction of binuclear iron center occurs upon oxygen uptake by hemerythrin results in production of bound peroxide (OOH-). Hemocyanin is most efficient in oxygen transport after hemoglobin. It contains binuclear copper (I) ion. Upon oxygenation O_2 reduced to peroxide (O_2^{2-}) consequently the two copper (I) atoms at the active site are oxidized to copper(II).

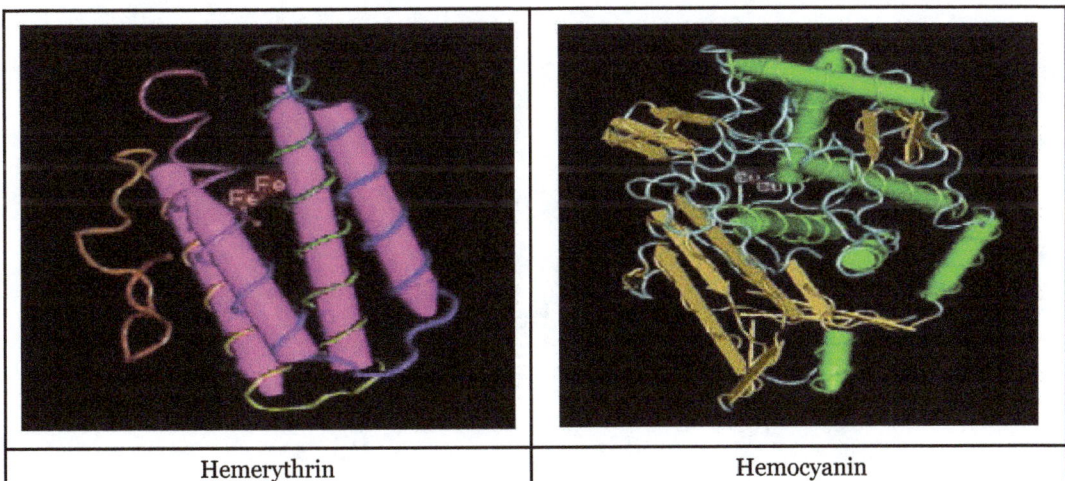

| Hemerythrin | Hemocyanin |

Electron Transfer

Cytochromes

Cytochromes are membrane bound heme containing proteins and are mainly responsible for ATP generation via electron transport. Cytochromes use redox behavior of
Fe^{2+}/Fe^{3+} which act as electron-transfer vectors. Cytochromes are thus, proficient in
performing oxidation and reduction reactions.

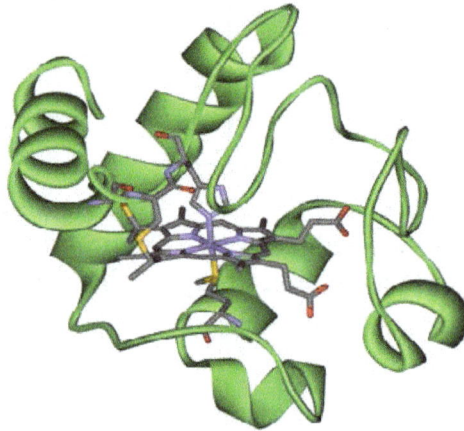

Cytochrome c

Moreover since the cytochromes are apprehended within membranes, thae redox reactions are carried out in the proper sequence for maximum efficiency. Most of the
cytochromes contained iron atom in a heme group. They differ in their side chains. For
example cytochrome a has a heme a prosthetic group and cytochrome b has a heme
b prosthetic group. Several cytochromes are involved in the mitochondrial electron
transport chain due to difference in Fe^{2+}/Fe^{3+} redox potentials arises from different
prosthetic group. Insertion of oxygen atom into C−H bond an oxidation reaction is
catalyse by cytochrome P450.

Rubredoxin

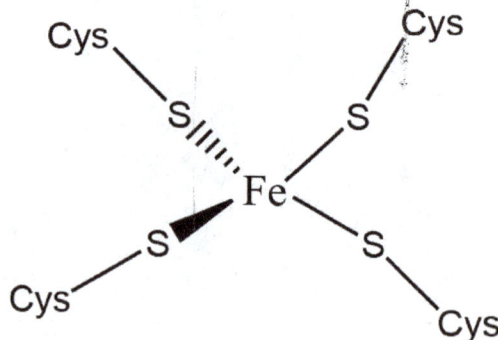

Rubredoxin active site

It is an electron-carrier protein found in sulfur-metabolizing bacteria and archaea. It governs one electron transfer processes. The active site of rubrdoxin consists of iron ion which is coordinated by the sulphur atoms of four cysteine residues in a tetrahedron arrangement. The oxidation state of central iron atom switches amidst the +2 and +3 oxidation states. The metal ion remains in high spin state in both the oxidation state, which minimizes any structural changes.

Plastocyanin

Plastocyanin belongs to "blue copper" proteins family which participates in electron transfer reactions. The preferred ligand geometry around copper atom is described as a "distorted trigonal pyramidal". Two nitrogen atoms of different histidines and a sulfur atom from cysteine forms the base of the pyramidal whereas methionine forms the apex by introducing an axial sulfur. The difference in bond length of two distinguished Cu-S bond causes rise in the redox potential of the protein. An absorption band appears at 597 nm due to the Cu-S bond, accounts for the blue color.

In the reduced form of plastocyanin, His-87 will become protonated with a pKa of 4.4. Protonation prevents it acting as a ligand and the copper site geometry becomes trigonal planar.

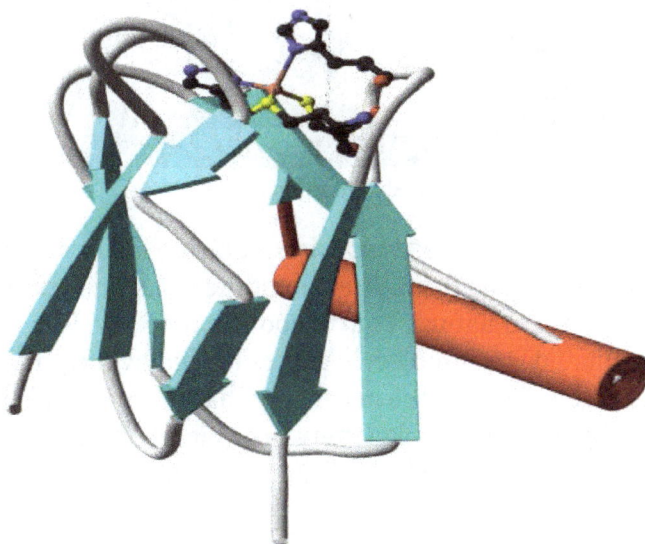

Plastocyanin (PDB ID: 3BQV)

Metalloenzymes

Metalloproteins with one labile coordination site around the metal centre are known as metalloenzyme. As with all enzymes, the shape of the active site is crucial. The metal ion is usually located in a pocket whose shape fits the substrate. The metal ion catalyzes reactions which are hard to achieve in organic chemistry.

Carbonic Anhydrase

In all living organisms, Carbon dioxide (CO_2) is a key metabolite. It exists in equilibrium with bicarbonate (HCO_3^-) poorly soluble in lipid membranes compared to carbon dioxide. Carbon dioxide can easily move in and out of the cell whereas bicarbonate could not. Thus *in vivo* a critical balance between conversion of bicarbonate to carbon dioxide and the reverse is required. The interconversion process of carbon dioxide and bicarbonate proceeds quite slowly at physiological pH which is catalyzed in presence of enzyme, Carbonic anhydrases. These are zinc-containing enzymes found in almost all living organisms; catalyse the reversible reaction between carbon dioxide hydration and bicarbonate dehydration. They play critical roles in assisting the transportation of carbon dioxide and protons in the intracellular space, across the biological membranes and also in the layers of the extracellular space. They take part in other processes, such as from respiration and photosynthesis in eukaryotes to cyanate degradation in the prokaryotes.

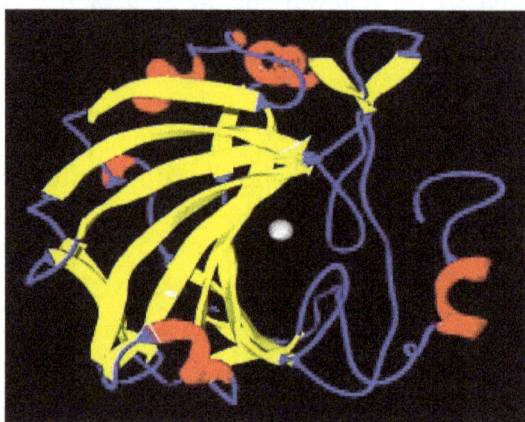

Carbonic anhydrase

The active site of carbonic anhydrases consists of a zinc ion, which is coordinated via three imidazole nitrogen atoms from three histidine units and coordination sphere is approximately tetrahedral in nature. Fourth coordination site is engaged with a water molecule. The coordinated water molecule gets polarized by the positively charged zinc ion and thus nucleophilic attack by the negatively charged hydroxide portion on carbon dioxide (carbonic anhydride) takes place very fast. Bicarbonate ion and hydrogen ions are produced in the catalytic cycle which remains in equilibrium.

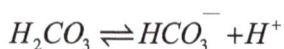

$$H_2CO_3 \rightleftharpoons HCO_3^- + H^+$$

Vitamin B$_{12}$-dependent Enzymes

Vitamin B$_{12}$ is known to catalyze transfer of methyl ($-CH_3$) groups between two molecules a process which is energetically expensive in organic reactions and it involves the breaking of C-C bonds. The activation energy required for the process is lowered due to presence of metal ion which forms a transient $Co-CH_3$ bond. The coenzyme consists of a cobalt(II) ion coordinated via four nitrogen atoms of a corrin rings and a fifth Nitrogen atom from an imidazole group. In the latent state, formation of a Co—C σ bond occurs with the 5' carbon atom of adenosine. This helps in understanding of its function in trans-methylation reactions as a naturally occurring organometallic compound such as the reaction carried out by methionine synthase.

Nitrogenase (Nitrogen Fixation)

The fixation of atmospheric nitrogen involves breaking very stable triple bond between the nitrogen atoms and thus requires intensive energy. This process is catalyzed by an enzyme known as nitrogenase, found in certain bacteria. There are three factors which enable it perform the necessary action and these are as follows: presence of a molybdenum atom at the active site, Iron-sulfur clusters engaged in transporting the electrons which are required to reduce the nitrogen along with a rich energy source. The nitrogen fixation can be written in the following manner

$$N_2 + 16MgATP + 8e^- \rightarrow 2NH_3 + 16MgADP + 16Pi + H_2$$

where Pi represents the inorganic phosphate. The active site contains a MoFe7S8 cluster which binds the dinitrogen molecule and allows the reduction process to start. The electron transportation occurs via associated "P" cluster which comprised of two cubical Fe4S4 clusters joined by sulfur bridges.

In vivo generation of the superoxide ion O2$^-$ occurs due to reduction of molecular oxygen. It behaves as a free radical due to presence of an unpaired electron and thus becomes a powerful oxidizing agent. These properties altogether enables the superoxide ion (very toxic) advantageous for the phagocytes to kill the harmful micro organisms. Else the superoxide ion must be damaged before it does unwanted cell damage. This function is delicately maintained by the superoxide dismutase enzymes.

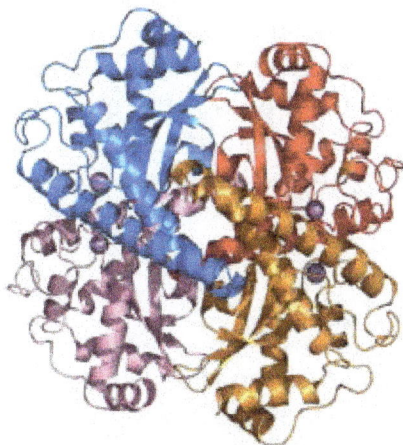

Superoxide dismutase

At neutral pH, in solution the superoxide ion disproportionate to molecular oxygen and hydrogen peroxide which is basically a dismutation reaction.

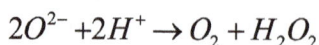

$$2O^{2-} + 2H^+ \rightarrow O_2 + H_2O_2$$

Both oxidation and reduction of superoxide ions are involved in a dismutation reaction. The superoxide dismutase group of enzymes is commonly known as SOD. A metal ion exhibiting varying oxidation state either act as an oxidizing agent or as a reducing agent is involved which is the central reason responsible for the action of these enzymes.

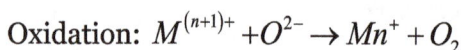

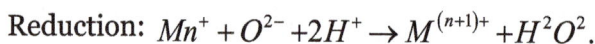

Oxidation: $M^{(n+1)+} + O^{2-} \rightarrow Mn^+ + O_2$

Reduction: $Mn^+ + O^{2-} + 2H^+ \rightarrow M^{(n+1)+} + H^2O^2$.

An example of this category is where the active metal is copper exists as Cu^{2+} or Cu^+ which is tetrahedrally coordinated by four histidine residues. Zinc ions in Human SOD are responsible for stabilization. Different other ions such as iron, manganese or nickel are found to be present in other isozymes. Another example which involves an unusual oxidation state of nickel that is nickel(III) is Ni-SOD. Ni geometry at the active site involves square planar Ni(II), thiolate (Cys2 and Cys6) and backbone nitrogen (His1 and Cys2) ligands and also square pyramidal Ni(III) with an added axial His1 side chain ligand.

Chlorophyll-containing Proteins

Chlorophyll exhibits a central role during photosynthesis. Chlorin ring absorbs photon due to its well defined electronic structure. It contains magnesium which is not directly involved in the photosynthetic function but can be replaced by other divalent ions with a slight reduction in activity.

In this process absorption of photon causes an electron to be excited into a singlet state which further undergoes an intersystem crossing from singlet state to triplet state. Thus it becomes a free radical (very reactive) and results an electron transfer to the acceptors which are adjacent to the chlorophyll in the chloroplast. In this phosynthetic cycle chlorophyll is oxidized and re-reduced. Finally molecular oxygen evolves as a final oxidation product as this process involves electrons withdrawal from water.

Signal-transduction

Metalloproteins Calmodulin

Calmodulin is an example of a signal-transduction protein. It contains four EF-hand (helix-loop- helix structural domain) motifs, which can binds Ca^{2+} ions. The ligating environment around calcium ion is adopts a pentagonal bipyramidal geometry. Calmodulin consist of two symmetrical globular domains (the N- and C-domain), separated by a by a flexible "hinge" region. The calcium ion acts as a diffusible second messenger to the initial stimuli and contributes in an intracellular signalling system.

Transcription Factors

The transcription factors are known to binds with DNA and are thus plays an important role in gene regulation. Zinc finger protein is known to be integral part of many transcription factors. Zinc finger is model where zinc ion at the centre of the protein surrounded by specific residues. Zinc finger proteins are known to binds with DNA but zinc ion does not directly participate in the process rather it is essential for the folding pattern of the protein. The ligating environment around the Zn^{2+} ion in these proteins usually have cysteine and histidine side chains.

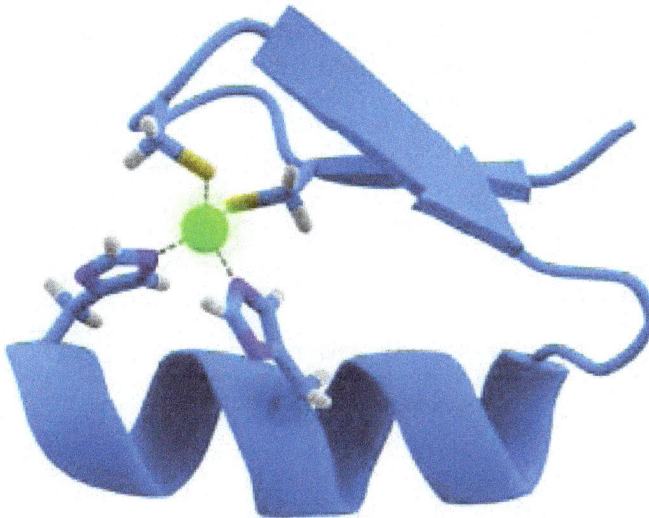

Zinc finger motif of proteins

Motor Proteins

The study of motor proteins has become a major focus in cell and molecular biology. Motor proteins are very interesting because they do what no man-made engines do: they directly convert the chemical energy into mechanical energy without using any electrical energy or heat as an intermediate. Motor proteins are enzymes that convert chemical energy into motion. Chemical energy is obtained from the hydrolysis of ATP and the motion is generated by the conformational changes depending on the bound nucleotide such as myosin, kinesin and dynein. Motor proteins play an important role in muscle contraction, cell migration, chromosome segregation, morphogenesis, beating of sperms and cilia, transport of intracellular cargoes etc.

Common Properties of Motor Proteins

- They move along the filaments.

- They can bind to specific filament types.

- They hydrolyze ATP.

Types of Motor Proteins

Generally there are two types of motor proteins-

(i) Actin based motor protein: Myosin is an actin based motor protein. It moves along the actin filaments.

(ii) Tubulin based motor protein: Tubulin is the building block of microtubules. Kinesin and dynein are tubulin based motor proteins. They move along the microtubule. Kinesin motor moves towards plus (+) end of microtubule (away from centrosome) whereas dynein moves towards minus (-) end of the micro-tubules (towards centrosome).

There are another type of motor proteins called Nucleic acid based motor proteins. They move along a DNA and produce force. DNA and RNA polymerases are nucleic acid based motor proteins.

Evolution of Motor Proteins

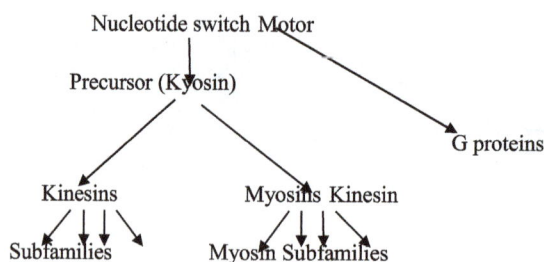

Examples of Kinesin Subfamilies	Direction	Processivity	Biological Activities
Conventional (Dimer)	Plus End	Yes	Membrane transport
Ncd (Dimer)	Minus End	No	Mitotic Spindle Function

Examples of Myosin Subfamilies

Myosin I (Monomer)	Barbed End	No	Cell motility/membrane functions
Myosin II (Dimer)	Barbed End	No	Muscle Contraction/Cytokinesis
Myosin V (Dimer)	Barbed End	Yes	Membrane/mRNA transport Myosin
VI (Dimer)	Pointed End	Unknown	Membrane Transport

Now before going to the details of the motor proteins we have to discuss about the structure of them.

Microtubules are hollow cylinders composed of α, β tubulin heteromers and actin is a helical polymer of 8 nm diameter. Actin is more flexible than microtubule. Both actin and microtubule are polar and the polarity is due to the fact that the individual subunits are asymmetric and they polymerize in head-to-tail manner.

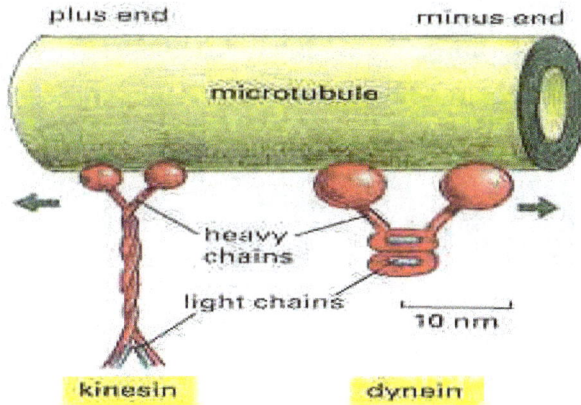

Actin Based Motility

Myosins are responsible for actin based motility. Myosins belong to a family of ATP dependent motor proteins. They play a crucial role in muscle contraction, vesicle transport etc. It mainly consists of head, neck and tail domain.

There are 18 different classes of myosin.

Myosin I contains one heavy chain with a single motor domain. It acts as monomer and functions in vesicle transport.

But Myosin II contains two heavy chains. The motor domain of each heavy chain has N terminal head domain and the C-terminal tail has coiled like morphology. Motor domain catalyzes the hydrolysis of ATP & it interacts with actin.

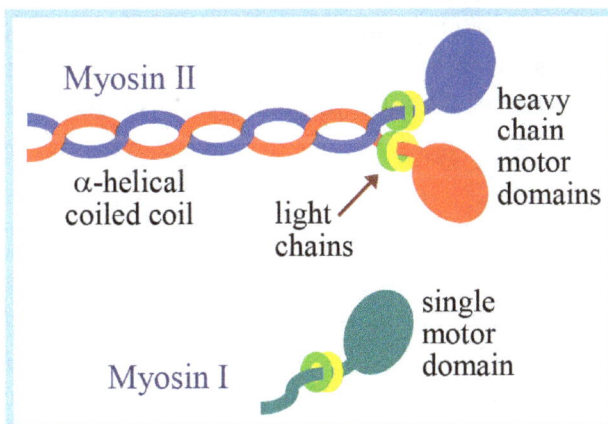

The structure of Myosin is given below:

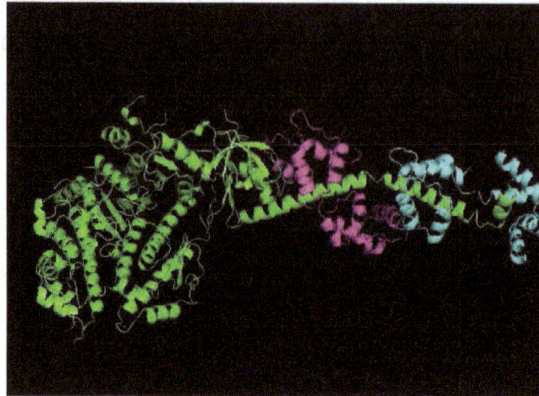

Myosin (PDB 1B7T)

How do Molecular Motor Works?

Myosin Cycle

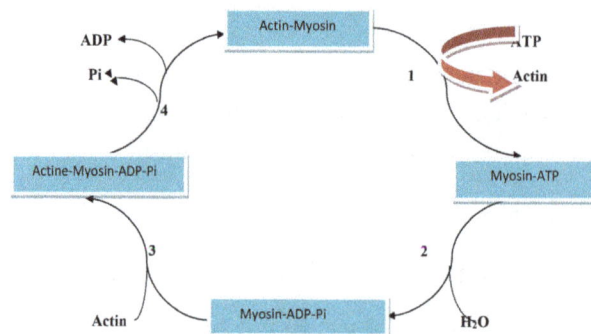

In the first step, ATP binds to myosin and induces the opening of actin binding. ATP binding causes a conformational change that causes myosin to let go off actin. ATP is then hydrolysed inducing a conformational change into high energy states (cocking), and these results in myosin weakly binding actin at a different place on the filament. Phosphate release causes myosin to bind very strongly with actin. Then ADP is released. ADP dissociation leaves the myosin head tightly bound to actin. Actin functions as nucleotide exchange factor.

Methods

There are several biophysical and structural approaches which have been applied to understand the mechanism of molecular motors. Here the aim is to identify all the intermediate conformational states during one ATP cycle. One of the suitable techniques is X-ray crystallography. This allows the visualization of protein 3D structure at the atomic resolution. But it is not possible to examine the motor track interaction with the help of X-ray crystallography. To make this possible electron microscopy has been introduced.

Studies of movement due to a single myosin molecule:

In this case a special type of setup is used where "optical traps" are created by focused laser beams. These optical traps can hold small objects and by adjusting the intensity of the laser beam the force can be controlled.

Actin filament is placed in optical trap via one or two attached beads.

Myosins are kept at low concentration so that only one myosin contacts the actin filament.

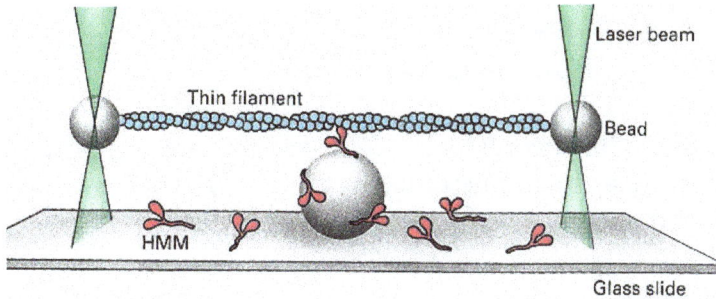

ATP is also kept low so that only one ATP binds to each myosin head.

In Vitro Motility Assays

In vitro motility assays provide an important approach to study the dynamics of motor protein movements. It is very useful to investigate myosin function using a small number of purified components. Now-a-days a special kind of in vitro assay has been developed in which fluorescently labelled actins filament were found to move over the glass surface coated with myosin. Using this assay it is possible to show that single headed myosin will support movement. In vitro motility assays can also measure the physical and mechanical properties of single molecule to establish the fundamental properties of these proteins.

There are two typical geometries that are used for in vitro motility assays: bead assays and surface assays. In case of bead assays, filaments are attached to a substrate, such as a microscope slide, and motors are attached to a small plastic bead, generally 1 μm in diameter, or to the tip of a fine glass needle. Light microscope helps to observe the motion of the beads or of the needle along the filaments in the presence of ATP. The position and movement of the beads or the needle are measured photo-electrically and can be determined with a resolution on the order of nanometres and sub-millisecond time-response. In case of surface assays, the motors themselves are attached to the substrate, and filaments are found to diffuse down from solution, attach to and glide over the motor-coated surface. This can be observed using dark-field or fluorescence microscopy.

In vitro motility assays. (a) The motile activity can be detected by attaching the motor to a bead and then allowing it to interact in an ATP-dependent manner with microtubules or oriented actin filaments on the cover glass surface.

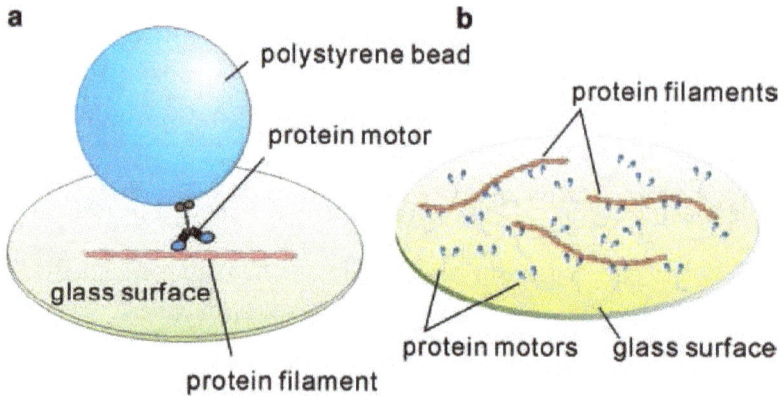

(b) Alternatively, individual fluorescence labelled actin filaments can be observed moving over a lawn of myosins and microtubules can be observed moving over a lawn of dynein or kinesin motors.

Molecular Motor

Molecular motors are biological molecular machines that are the essential agents of movement in living organisms. In general terms, a motor is a device that consumes energy in one form and converts it into motion or mechanical work; for example, many protein-based molecular motors harness the chemical free energy released by the hydrolysis of ATP in order to perform mechanical work. In terms of energetic efficiency, this type of motor can be superior to currently available man-made motors. One important difference between molecular motors and macroscopic motors is that molecular motors operate in the thermal bath, an environment in which the fluctuations due to thermal noise are significant.

Examples

Some examples of biologically important molecular motors:

- Cytoskeletal motors

  o Myosins are responsible for muscle contraction, intracellular cargo transport, and producing cellular tension.

  o Kinesin moves cargo inside cells away from the nucleus along microtubules.

  o Dynein produces the axonemal beating of cilia and flagella and also transports cargo along microtubules towards the cell nucleus.

- Polymerisation motors

  o Actin polymerization generates forces and can be used for propulsion. ATP is used.

- o Microtubule polymerization using GTP.

- o Dynamin is responsible for the separation of clathrin buds from the plasma membrane. GTP is used.

- Rotary motors:

  - o F_oF_1-ATP synthase family of proteins convert the chemical energy in ATP to the electrochemical potential energy of a proton gradient across a membrane or the other way around. The catalysis of the chemical reaction and the movement of protons are coupled to each other via the mechanical rotation of parts of the complex. This is involved in ATP synthesis in the mitochondria and chloroplasts as well as in pumping of protons across the vacuolar membrane.

  - o The bacterial flagellum responsible for the swimming and tumbling of *E. coli* and other bacteria acts as a rigid propeller that is powered by a rotary motor. This motor is driven by the flow of protons across a membrane, possibly using a similar mechanism to that found in the F_o motor in ATP synthase.

- Nucleic acid motors:

  - o RNA polymerase transcribes RNA from a DNA template.

  - o DNA polymerase turns single-stranded DNA into double-stranded DNA.

  - o Helicases separate double strands of nucleic acids prior to transcription or replication. ATP is used.

  - o Topoisomerases reduce supercoiling of DNA in the cell. ATP is used.

  - o RSC and SWI/SNF complexes remodel chromatin in eukaryotic cells. ATP is used.

  - o SMC proteins responsible for chromosome condensation in eukaryotic cells.

  - o Viral DNA packaging motors inject viral genomic DNA into capsids as part of their replication cycle, packing it very tightly. Several models have been put forward to explain how the protein generates the force required to drive the DNA into the capsid. An alternative proposal is that, in contrast with all other biological motors, the force is not generated directly by the protein, but by the DNA itself. In this model, ATP hydrolysis is used to drive protein conformational changes that alternatively dehydrate and rehydrate the DNA, cyclically driving it from B-DNA to A-DNA and back again. A-DNA is 23% shorter than B-DNA, and the DNA shrink/expand cycle is coupled to a protein-DNA grip/release cycle to generate the forward motion that propels DNA into the capsid.

- Synthetic molecular motors have been created by chemists that yield rotation, possibly generating torque.

Theoretical Considerations

Because the motor events are stochastic, molecular motors are often modeled with the Fokker-Planck equation or with Monte Carlo methods. These theoretical models are especially useful when treating the molecular motor as a Brownian motor.

Experimental Observation

In experimental biophysics, the activity of molecular motors is observed with many different experimental approaches, among them:

- Fluorescent methods: fluorescence resonance energy transfer (FRET), fluorescence correlation spectroscopy (FCS), total internal reflection fluorescence (TIRF).

- Magnetic tweezers can also be useful for analysis of motors that operate on long pieces of DNA.

- Neutron spin echo spectroscopy can be used to observe motion on nanosecond timescales.

- Optical tweezers (not to be confused with molecular tweezers in context) are well-suited for studying molecular motors because of their low spring constants.

- Scattering techniques: single particle tracking based on dark field microscopy or interferometric scattering microscopy (iSCAT)

- Single-molecule electrophysiology can be used to measure the dynamics of individual ion channels.

Many more techniques are also used. As new technologies and methods are developed, it is expected that knowledge of naturally occurring molecular motors will be helpful in constructing synthetic nanoscale motors.

Non-biological

Recently, chemists and those involved in nanotechnology have begun to explore the possibility of creating molecular motors *de novo*. These synthetic molecular motors currently suffer many limitations that confine their use to the research laboratory. However, many of these limitations may be overcome as our understanding of chemistry and physics at the nanoscale increases. Systems like the nanocars, while not technically motors, are illustrative of recent efforts towards synthetic nanoscale motors.

Now we will focus on tubulin based motor proteins.

Tubulin Based Motility

Kinesins and Dyneins are responsible for tubulin based motility.

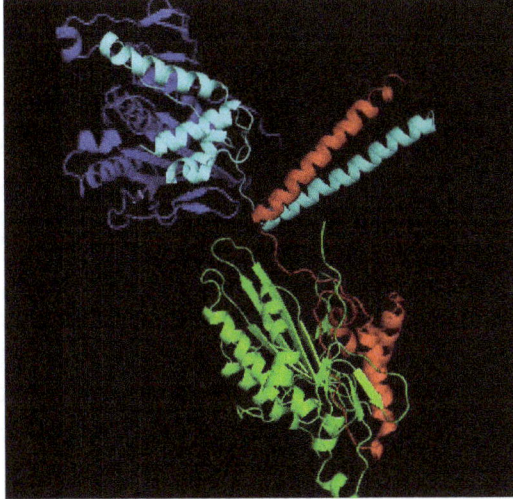

Kinesin (PDB 3KIN)

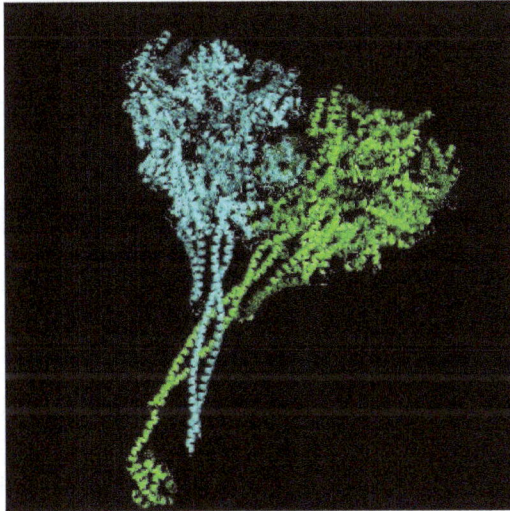

Dynein (PDB 3VKH)

There are 16 different classes of kinesin. Kinesin is mainly found in eukaryotic cell. It contains two motor domains that are joined through coiled-coiled stalk. ATP binds to the motor protein and the free energy required for the motility is supplied by ATP. Motor domain can also binds to the microtubules in an ATP dependent fashion.

The heavy chains of dynein contain C-terminal head with two elongated flexible structures called the stalk (microtubule-binding domain) and the N-terminal tail (cargo-binding domain). The head and the stalk form a motor domain. Dyneins can be divided into two groups: cytoplasmic dyneins and axonemal dyneins. Cytoplasmic dyneins play a crucial role in the transport of cargos such as vesicles whereas Axonemal dyneins play a major role in the movement of cilia and flagella.

Cytoplasm

In cell biology, the cytoplasm is the material within a living cell, excluding the cell nucleus. It comprises cytosol (the gel-like substance enclosed within the cell membrane) and the organelles – the cell's internal sub-structures. All of the contents of the cells of prokaryotic organisms (such as bacteria, which lack a cell nucleus) are contained within the cytoplasm. Within the cells of eukaryotic organisms the contents of the cell nucleus are separated from the cytoplasm, and are then called the nucleoplasm. The cytoplasm is about 80% water and usually colorless.

The submicroscopic ground cell substance or cytoplasmatic matrix which remains after exclusion the cell organelles and particles is groundplasm. It is the hyaloplasm of light microscopy, and high complex, polyphasic system in which all of resolvable cytoplasmic elements of are suspended, including the larger organelles such the ribosomes, mitochondria, the plant plastids, lipid droplets, and vacuoles.

It is within the cytoplasm that most cellular activities occur, such as many metabolic pathways including glycolysis, and processes such as cell division. The concentrated inner area is called the endoplasm and the outer layer is called the cell cortex or the ectoplasm.

Movement of calcium ions in and out of the cytoplasm is a signaling activity for metabolic processes.

In plants, movement of the cytoplasm around vacuoles is known as cytoplasmic streaming.

History

The term was introduced by Rudolf von Kölliker in 1863, originally as a synonym for protoplasm, but later it has come to mean the cell substance and organelles outside the nucleus.

There has been certain disagreement on the definition of cytoplasm, as some authors prefer to exclude from it some organelles, especially the vacuoles and sometimes the plastids.

Physical Nature

The physical properties of the cytoplasm have been contested in recent years. It remains uncertain how the varied components of the cytoplasm interact to allow movement of particles and organelles while maintaining the cell's structure. The flow of cytoplasmic components plays an important role in many cellular functions which are dependent on the permeability of the cytoplasm. An example of such function is cell signalling, a process which is dependent on the manner in which signaling molecules are allowed to diffuse across the cell. While small signaling molecules like calcium ions are able to diffuse with ease, larger molecules and sub-cellular structures often require aid in moving through the cytoplasm. The irregular dynamics of such particles have given rise to various theories on the nature of the cytoplasm.

As a Sol-gel

There has long been evidence that the cytoplasm behaves like a sol-gel. It is thought that the component molecules and structures of the cytoplasm behave at times like a disordered colloidal solution (sol) and at other times like an integrated network, forming a solid mass (gel). This theory thus proposes that the cytoplasm exists in distinct fluid and solid phases depending on the level of interaction between cytoplasmic components, which may explain the differential dynamics of different particles observed moving through the cytoplasm.

As a Glass

Recently it has been proposed that the cytoplasm behaves like a glass-forming liquid approaching the glass transition. In this theory, the greater the concentration of cytoplasmic components, the less the cytoplasm behaves like a liquid and the more it behaves as a solid glass, freezing larger cytoplasmic components in place (it is thought that the cell's metabolic activity is able to fluidize the cytoplasm to allow the movement of such larger cytoplasmic components). A cell's ability to vitrify in the absence of metabolic activity, as in dormant periods, may be beneficial as a defence strategy. A solid glass cytoplasm would freeze subcellular structures in place, preventing damage, while allowing the transmission of very small proteins and metabolites, helping to kickstart growth upon the cell's revival from dormancy.

Other Perspectives

There has been research examining the motion of cytoplasmic particles independent of the nature of the cytoplasm. In such an alternative approach, the aggregate random forces within the cell caused by motor proteins explain the non-Brownian motion of cytoplasmic constituents.

Constituents

The three major elements of the cytoplasm are the cytosol, organelles and inclusions.

Cytosol

The cytosol is the portion of the cytoplasm not contained within membrane-bound organelles. Cytosol makes up about 70% of the cell volume and is a complex mixture of cytoskeleton filaments, dissolved molecules, and water. The cytosol's filaments include the protein filaments such as actin filaments and microtubules that make up the cytoskeleton, as well as soluble proteins and small structures such as ribosomes, proteasomes, and the mysterious vault complexes. The inner, granular and more fluid portion of the cytoplasm is referred to as endoplasm.

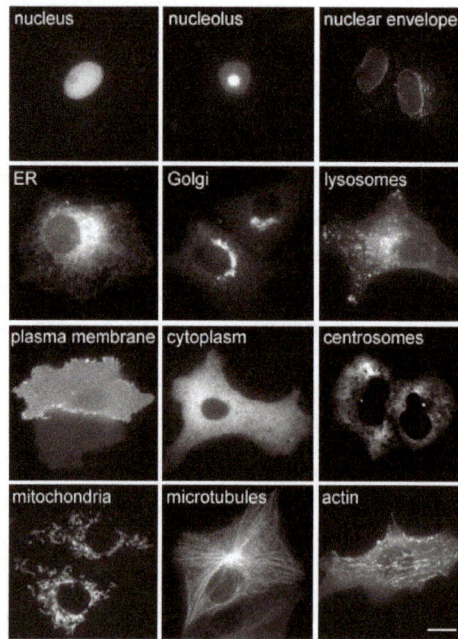

with friendly permission of Jeremy Simpson and Rainer Pepperkok

Proteins in different cellular compartments and structures tagged with green fluorescent protein

Due to this network of fibres and high concentrations of dissolved macromolecules, such as proteins, an effect called macromolecular crowding occurs and the cytosol does not act as an ideal solution. This crowding effect alters how the components of the cytosol interact with each other.

Organelles

Organelles (literally "little organs"), are usually membrane-bound structures inside the cell that have specific functions. Some major organelles that are suspended in the cytosol are the mitochondria, the endoplasmic reticulum, the Golgi apparatus, vacuoles, lysosomes, and in plant cells, chloroplasts.

Cytoplasmic Inclusions

The inclusions are small particles of insoluble substances suspended in the cytosol. A huge range of inclusions exist in different cell types, and range from crystals of calcium oxalate or silicon dioxide in plants, to granules of energy-storage materials such as starch, glycogen, or polyhydroxybutyrate. A particularly widespread example are lipid droplets, which are spherical droplets composed of lipids and proteins that are used in both prokaryotes and eukaryotes as a way of storing lipids such as fatty acids and sterols. Lipid droplets make up much of the volume of adipocytes, which are specialized lipid-storage cells, but they are also found in a range of other cell types.

Controversy and Research

The cytoplasm, mitochondria and most organelles are contributions to the cell from the maternal gamete. Contrary to the older information that disregards any notion of the cytoplasm being active, new research has shown it to be in control of movement and flow of nutrients in and out of the cell by viscoplastic behavior and a measure of the reciprocal rate of bond breakage within the cytoplasmic network.

The material properties of the cytoplasm remain an ongoing investigation. Recent measurements using force spectrum microscopy reveal that the cytoplasm can be likened to an elastic solid, rather than a viscoelastic fluid.

Axoneme

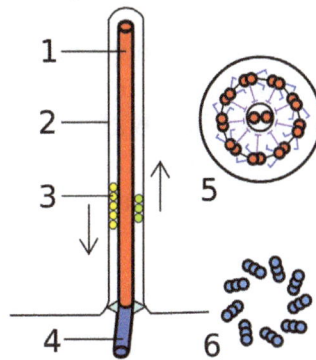

Eukaryotic flagella. 1-axoneme, 2-cell membrane, 3-IFT (intraflagellar transport), 4-basal body, 5-cross section of flagella, 6-triplets of microtubules of basal body

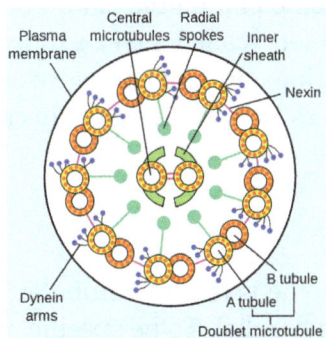

Cross section of an axoneme

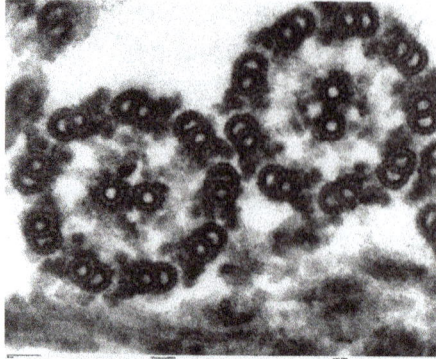

Micrograph of thin x-section cut through *Chlamydomonas* axoneme

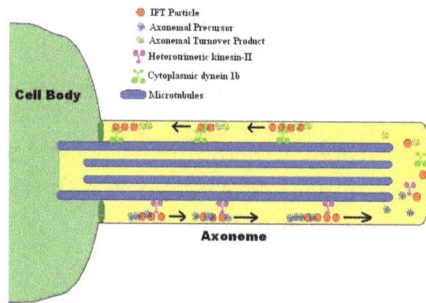

A simplified model of intraflagellar transport

Numerous eukaryotic cells carry whip-like appendages (cilia or eukaryotic flagella) whose inner core consists of a cytoskeletal structure called the axoneme. The axoneme serves as the "skeleton" of these organelles, both giving support to the structure and, in some cases, causing it to bend. Though distinctions of function and/or length may be made between cilia and flagella, the internal structure of the axoneme is common to both.

Structure

Inside cilia and flagella is a microtubule-based cytoskeleton called the axoneme. The axoneme of primary cilia typically has a ring of nine outer microtubule triplets (called a 9+0 axoneme), and the axoneme of a motile cilium has two central microtubules in addition to the nine outer doublets (called a 9+2 axoneme). The axonemal cytoskeleton acts as a scaffolding for various protein complexes and provides binding sites for molecular motor proteins such as kinesin II, that help carry proteins up and down the microtubules.

Motile Cilia

The building-block of the axoneme is the microtubule; each axoneme is composed of several microtubules aligned in parallel. To be specific, the microtubules are arranged in a characteristic pattern known as the "9x2 + 2," as shown in the image at right. Nine sets of "doublet" microtubules (a specialized structure consisting of two linked microtubules) form a ring around a "central pair" of single microtubules.

Besides the microtubules, the axoneme contains many proteins and protein complexes necessary for its function. The dynein arms, for example, are motor complexes that produce the force needed for bending. Each dynein arm is anchored to a doublet microtubule; by "walking" along an adjacent microtubule, the dynein motors can cause the microtubules to slide against each other. When this is carried out in a synchronized fashion, with the microtubules on one side of the axoneme being pulled 'down' and those on the other side pulled 'up,' the axoneme as a whole can bend back and forth. This process is responsible for ciliary/flagellar beating, as in the well-known example of the human sperm.

The radial spoke is another protein complex of the axoneme. Thought to be important in regulating the motion of the axoneme, this "T"-shape complex projects from each set of outer doublets toward the central microtubules. The inter-doublet connections between adjacent microtubule pairs are termed nexin linkages.

Non-motile/primary Cilia

The axoneme structure in non-motile ("primary") cilia shows some variation from the canonical "9x2 + 2" anatomy. No dynein arms are found on the outer doublet microtubules, and there is no pair of central microtubule singlets. This organization of axoneme is referred as "9x2 + 0". In addition, "9x2 + 1" axonemes, with only a single central microtubule, have been found to exist. Primary cilia appear to serve sensory functions.

Clinical Significance

Mutations or defects in primary cilia have been found to play a role in human diseases. These ciliopathies include polycystic kidney disease (PKD), retinitis pigmentosa, Bardet-Biedl syndrome, and other developmental defects.

Processivity & Duty Ratio

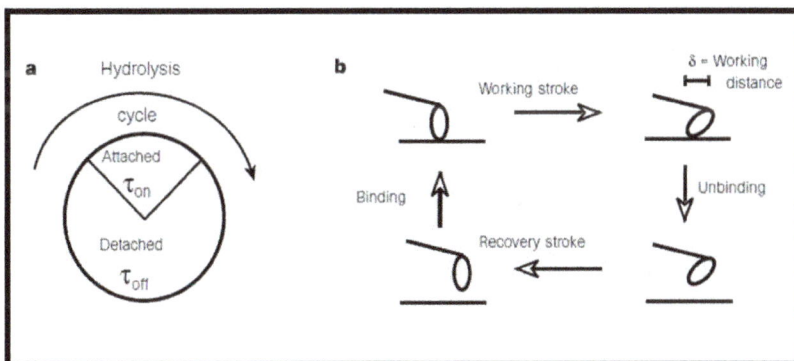

Duty ratio (r) = $\delta V/v$ where, δ = working distance or step size

V = ATPase activity

v = in vitro motility velocity

r = duty ratio

Processive motor, r->1

Non processive motor, r->0

Kinesin is coordinated but ATP cycle in both heads is out of phase. Here one head is always connected to microtubule. Therefore kinesin is highly processive.

In case of myosin II the two heads are not coordinated and cycle in both heads is independent of each other. Therefore myosin II is not processive.

Kinesin Cycle

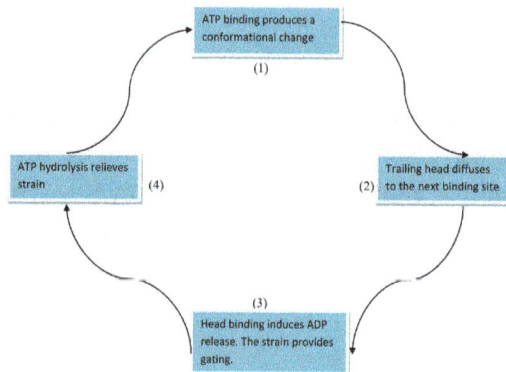

Kinesin is highly processive. To explain the mechanism of processive movement of kinesin several models have been proposed-

- Hand-over-hand model
- Inchworm model
- Thermal ratchet model

Hand-over-hand Model

Strong microtubule binding takes place at ATP-bound state whereas weak microtubule binding takes place at A DP-bound state. Kinesin is a double headed motor protein and exchange of leading head and trailing head occurs in this model. Here the two heads walk in a hand-over-hand manner i.e. in a coordinated manner.

In the hand-over-hand model, the binding of hydrolysis of an ATP generates a conformational change in the forward head (head 1) and this conformation pulls the rear head

(head 2) forward, while head 1 remains fixed. In the next step, head 2 remains fixed and forces the head 1 forward. Binding of ATP to leading head induces power stroke.

The hand-over-hand model has several advantages.

First, a single-headed molecule should not be processive.

Secondly, though a single-headed molecule does not move, it should have a high affinity for the microtubule. This is due to the absence of second head to induce its detachment.

Third, we might expect that single-headed molecules move very slowly because there is a loss of mechanical amplification provided by the other head such as myosin which moves more slowly when it acts as a mechanical lever.

Although the hand-over-hand model is attractive, it has some disadvantages too.

First, the atomic structure of dimeric kinesin with bound ADP shows that the two heads cannot simultaneously bind to the polar microtubule unless the dimer is highly strained and they are rotated by ~120° with respect to each other. This indicates that the motion is purely one headed and the second head is a passenger which has no active role.

Second, there are kinesin-related proteins, KIF1A and KIF1B. Like conventional kinesin, they are thought to be vesicle and organelle transporters, but they are monomeric. The existence of these motors shows that a double-headed structure is not necessary for organelle motility, and from these a question also came into the mind that how monomeric motors could maintain contact with the microtubule when they move.

Inchworm model: Here one kinesin head always lead & one of the two heads hydrolyzes ATP. The exchange of leading head and trailing head does not occur in this model. One-headed kinesin has an ordered mechanism where Pi is released before ADP. The following evidences support these. In the absence of microtubules both the monomeric and dimeric kinesin releases Pi before ADP whereas in case of myosin motor domains they release Pi before ADP both in presence & absence of actin.

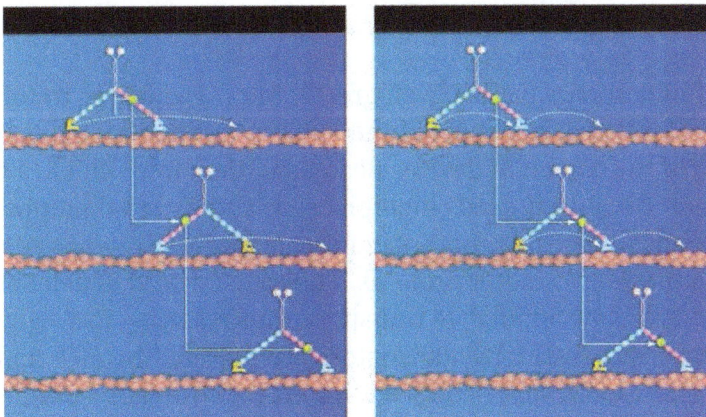

Hand-over-hand model Inchworm model

Thermal Ratchet Model

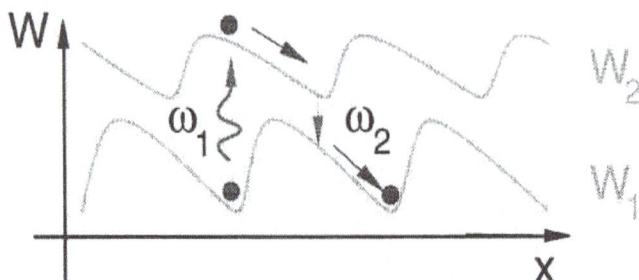

The motor protein is described as a particle that moves in a periodic manner consisting of a series of asymmetric potential wells. Due to ATP hydrolysis the well depths fluctuate in time. This leads to the formation of a net particle flux which is in the direction toward the steeper part of the potentials.

Advantages

    1) Continuum description well developed formalism.

    2) Convenient for numerical calculations and simulations.

    3) Small number of parameters.

Disadvantages

    1) Mainly numerical or simulations results.

    2) Results depend on potentials used in calculations.

    3) Difficult to make quantitative comparisons with experiments.

    4) Not flexible in description of complex biochemical networks.

Molecular Motors for Advanced Technology Devices

The efficiency of immunoassays has been greatly improved by microfluidics. Microfluidics can be used for active transport of the analyte through a small detection device. If the dimension of the detection devices decreases, then microfluidics cannot be used for active transport because it needs high pressure and external pumps. This is where motor proteins come in. The transport of nanoscale cargoes in nanoscale environments can be done with the help of motor proteins. They can be used in detection devices and can replace the fluidic flow for analyte transport, thus resulting in a significant increase in efficiency and sensitivity. Molecular motors are powered by ATP and the energy of the detection device can be supplied by adding ATP to the sample buffer. These make the device completely independent of external power sources. This leads to lab-on-a-chip detection devices that are easy to handle like a urine test strip or a home-use pregnancy test.

Kinesin Defects

- Some neurodegenerative diseases arise from the defects in kinesin-mediated long distance transport of materials along microtubules.

- Some types of cancers are also associated with abnormalities of kinesins involved in mitosis.

- Deficiency of kinesins involved in transport of cargo to the tips of cilia and flagella can also lead to the formation of Ciliary and flagellar defects.

Membrane Proteins

Proteins consist of three main classes which are classified as globular, fibrous and membrane proteins. A cell is enveloped by a membrane which makes the boundary of a cell and enables it to maintain the distinction between cytosolic and extracellular milieu. Cells consist of various organelles such as golgi body, endoplasmic reticulum, mitochondria and several other membrane bound organelles. The difference between cytosol and these organelles are maintained by individual membranes. These biological membranes are made up of mainly lipid bilayers whereas functions are carried out by membrane proteins.

Classification of Membrane Proteins

Membrane associated proteins can be classified in the following two ways:

- Mode of interaction with the membranes
- Cellular locations

According to the literature, the membrane proteins can be categorized as follows:

(1) Type I membrane proteins

(2) Type II membrane proteins

(3) Multipass transmembrane proteins

(4) Lipid chain anchored-membrane proteins

(5) GPI-anchored-membrane proteins

(6) Peripheral membrane proteins

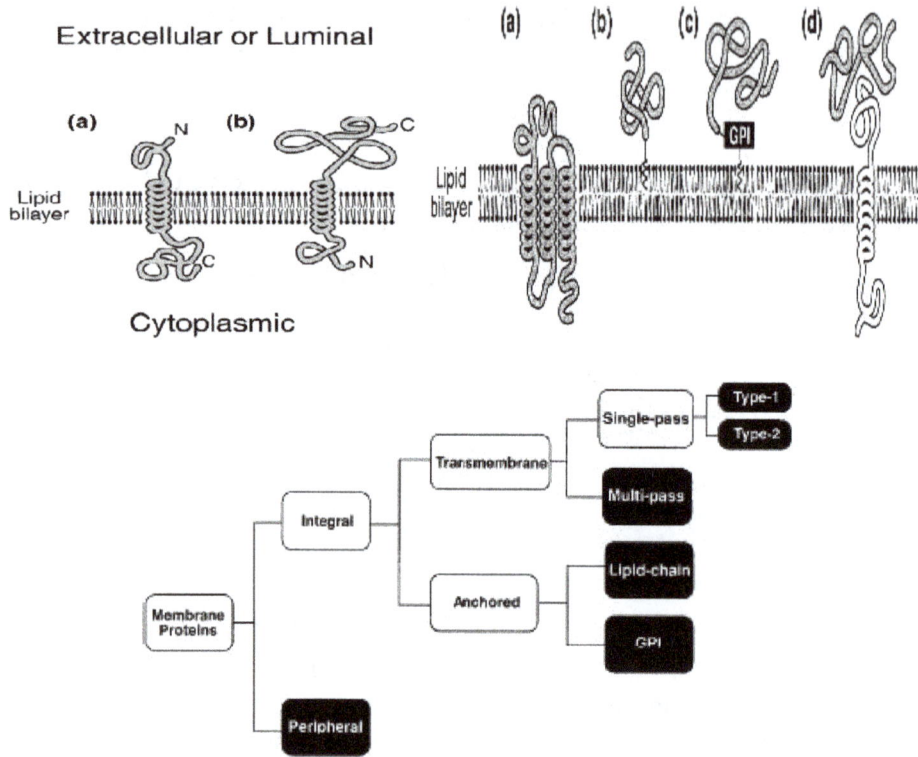

Extracellular or Luminal

Cytoplasmic

Fig. 3. Categorization chart to show the relationship of the six membrane protein types illustrated in Figs. 1 and 2.

Integral Membrane Protein

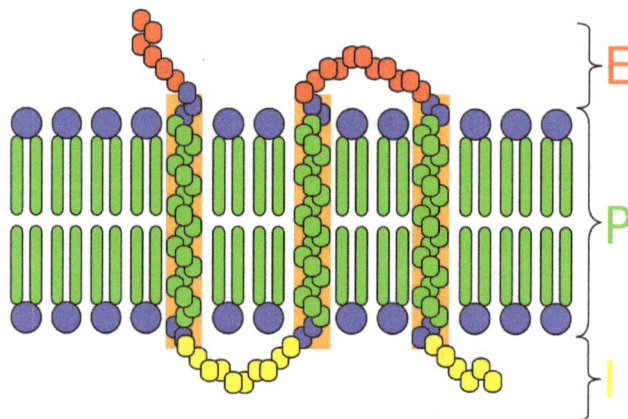

E=extracellular space; P=plasma membrane; I=intracellular space

An integral membrane protein (IMP) is a type of membrane protein that is permanently attached to the biological membrane. All transmembrane proteins are IMPs, but not all IMPs are transmembrane proteins. IMPs comprise a significant fraction of the proteins encoded in an organism's genome. Proteins that cross the membrane are surrounded by "annular" lipids, which are defined as lipids that are in direct contact with a membrane protein. Such proteins can be separated from the biological membranes only using detergents, nonpolar solvents, or sometimes denaturing agents.

Structure

Three-dimensional structures of ~160 different integral membrane proteins have been determined at atomic resolution by X-ray crystallography or nuclear magnetic resonance spectroscopy. They are challenging subjects for study owing to the difficulties associated with extraction and crystallization. In addition, structures of many water-soluble protein domains of IMPs are available in the Protein Data Bank. Their membrane-anchoring α-helices have been removed to facilitate the extraction and crystallization. Search integral membrane proteins in the PDB (based on gene ontology classification).

IMPs can be divided into two groups:

1. Integral polytopic proteins (Transmembrane proteins)

2. Integral monotopic proteins

Integral Polytopic Protein

The most common type of IMP is the transmembrane protein (TM), which spans the entire biological membrane. Single-pass membrane proteins cross the membrane only once, while multi-pass membrane proteins weave in and out, crossing several times. Single pass TM proteins can be categorized as Type I, which are positioned such that their carboxyl-terminus is towards the cytosol, or Type II, which have their amino-terminus towards the cytosol. Type III proteins have multiple transmembrane domains in a single polypeptide, while type IV consists of several different polypeptides assembled together in a channel through the membrane. Type V proteins are anchored to the lipid bilayer through covalently linked lipids. Finally Type VI proteins have both a transmembrane domains and lipid anchors.

Integral Monotopic Proteins

Integral monotopic proteins are associated to the membrane from one side but do not span the lipid bilayer completely.

Determination of Protein Structure

The Protein Structure Initiative (PSI), funded by the U.S. National Institute of General Medical Sciences (NIGMS), part of the National Institutes of Health (NIH), has among its aim to determine three-dimensional protein structures and to develop techniques for use in structural biology, including for membrane proteins. Homology modeling can be used to construct an atomic-resolution model of the "target" integral protein from its amino acid sequence and an experimental three-dimensional structure of a related homologous protein. This procedure has been extensively used for ligand-G protein–coupled receptors (GPCR) and their complexes.

Function

IMPs include transporters, linkers, channels, receptors, enzymes, structural membrane-anchoring domains, proteins involved in accumulation and transduction of energy, and proteins responsible for cell adhesion. Classification of transporters can be found in Transporter Classification Database.

In this case the integral membrane protein spans the phospholipid bilayer seven times. The part of the protein that is embedded in the hydrophobic regions of the bilayer are alpha helical and composed of predominantly hydrophobic amino acids. The amino terminal end of the protein is in the cytosol while the C terminal region is in the outside of the cell. A membrane that contains this particular protein is able to function in photosynthesis.

Peripheral Membrane Protein

Peripheral membrane proteins are membrane proteins that adhere only temporarily to the biological membrane with which they are associated. These proteins attach to integral membrane proteins, or penetrate the peripheral regions of the lipid bilayer. The regulatory protein subunits of many ion channels and transmembrane receptors, for example, may be defined as peripheral membrane proteins. In contrast to integral membrane proteins, peripheral membrane proteins tend to collect in the water-soluble component, or fraction, of all the proteins extracted during a protein purification procedure. Proteins with GPI anchors are an exception to this rule and can have purification properties similar to those of integral membrane proteins.

The reversible attachment of proteins to biological membranes has shown to regulate cell signaling and many other important cellular events, through a variety of mechanisms. For example, the close association between many enzymes and biological membranes may bring them into close proximity with their lipid substrate(s). Membrane binding may also promote rearrangement, dissociation, or conformational changes within many protein structural domains, resulting in an activation of their biological activity. Additionally, the positioning of many proteins are localized to either the inner or outer surfaces or leaflets of their resident membrane. This facilitates the assembly of multi-protein complexes by increasing the probability of any appropriate protein–protein interactions.

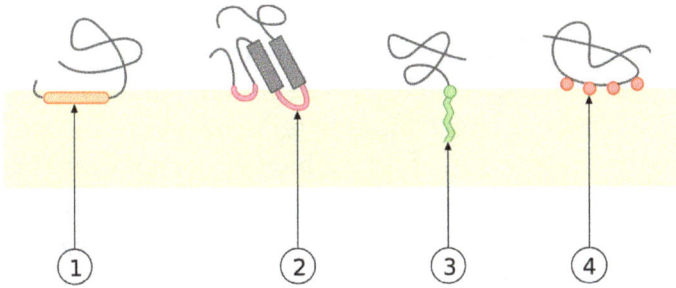

Schematic representation of the different types of interaction between monotopic membrane proteins and the cell membrane: 1. interaction by an amphipathic α-helix parallel to the membrane plane (in-plane membrane helix) 2. interaction by a hydrophobic loop 3. interaction by a covalently bound membrane lipid (*lipidation*) 4. electrostatic or ionic interactions with membrane lipids (*e.g.* through a calcium ion)

Binding to the Lipid Bilayer

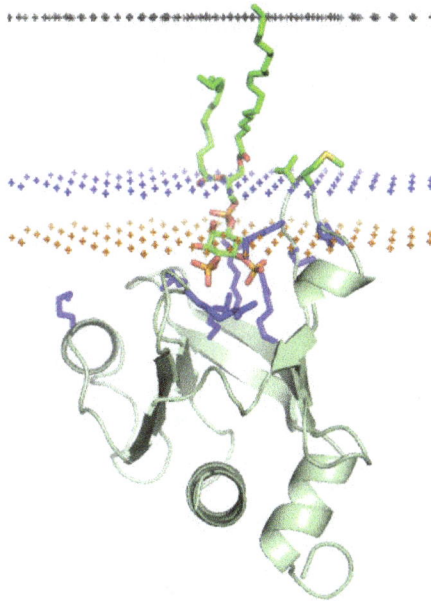

PH domain of phospholipase C delta 1. Middle plane of the lipid bilayer – black dots. Boundary of the hydrocarbon core region – blue dots (intracellular side). Layer of lipid phosphates – yellow dots.

Peripheral membrane proteins may interact with other proteins or directly with the lipid bilayer. In the latter case, they are then known as *amphitropic* proteins. Some proteins, such as G-proteins and certain protein kinases, interact with transmembrane proteins and the lipid bilayer simultaneously. Some polypeptide hormones, antimicrobial peptides, and neurotoxins accumulate at the membrane surface prior to locating and interacting with their cell surface receptor targets, which may themselves be peripheral membrane proteins.

The phospholipid bilayer that forms the cell surface membrane consists of a hydrophobic inner core region sandwiched between two regions of hydrophilicity, one at the inner surface and one at the outer surface of the cell membrane. The inner and outer surfaces, or interfacial regions, of model phospholipid bilayers have been shown to have a thickness of around 8 to 10 Å, although this may be wider in biological membranes that include large amounts of gangliosides or lipopolysaccharides. The hydrophobic inner core region of typical biological membranes may have a thickness of around 27 to 32 Å, as estimated by Small angle X-ray scattering (SAXS). The boundary region between the hydrophobic inner core and the hydrophilic interfacial regions is very narrow, at around 3Å, (see lipid bilayer article for a description of its component chemical groups). Moving outwards away from the hydrophobic core region and into the interfacial hydrophilic region, the effective concentration of water rapidly changes across this boundary layer, from nearly zero to a concentration of around 2 M. The phosphate groups within phospholipid bilayers are fully hydrated or saturated with water and are situated around 5 Å outside the boundary of the hydrophobic core region.

Some water-soluble proteins associate with lipid bilayers *irreversibly* and can form transmembrane alpha-helical or beta-barrel channels. Such transformations occur in pore forming toxins such as colicin A, alpha-hemolysin, and others. They may also occur in BcL-2 like protein , in some amphiphilic antimicrobial peptides , and in certain annexins . These proteins are usually described as peripheral as one of their conformational states is water-soluble or only loosely associated with a membrane.

Membrane Binding Mechanisms

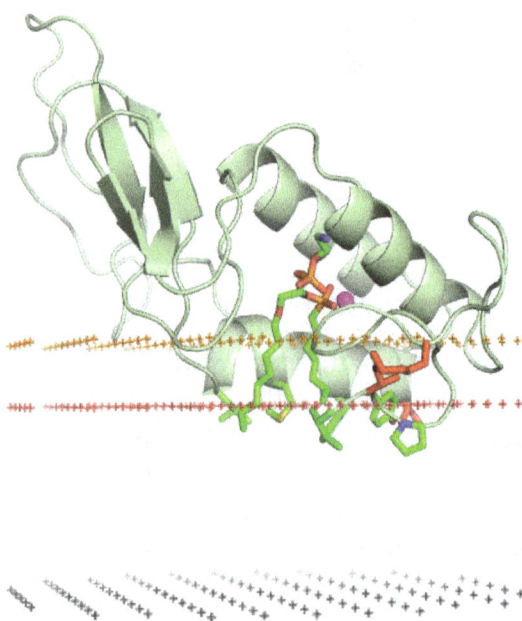

Bee venom phospholipase A2 (1poc). Middle plane of the lipid bilayer – black dots. Boundary of the hydrocarbon core region – red dots (extracellular side). Layer of lipid phosphates – yellow dots.

The association of a protein with a lipid bilayer may involve significant changes within tertiary structure of a protein. These may include the folding of regions of protein structure that were previously unfolded or a re-arrangement in the folding or a refolding of the membrane-associated part of the proteins . It also may involve the formation or dissociation of protein quaternary structures or oligomeric complexes, and specific binding of ions, ligands, or regulatory lipids.

Typical amphitropic proteins must interact strongly with the lipid bilayer in order to perform their biological functions. These include the enzymatic processing of lipids and other hydrophobic substances, membrane anchoring, and the binding and transfer of small nonpolar compounds between different cellular membranes. These proteins may be anchored to the bilayer as a result of hydrophobic interactions between the bilayer and exposed nonpolar residues at the surface of a protein, by specific non-covalent binding interactions with regulatory lipids , or through their attachment to covalently bound lipid anchors.

It has been shown that the membrane binding affinities of many peripheral proteins depend on the specific lipid composition of the membrane with which they are associated.

Non-specific Hydrophobic Association

Amphitropic proteins associate with lipid bilayers via various hydrophobic anchor structures. Such as amphiphilic α-helices, exposed nonpolar loops, post-translationally acylated or lipidated amino acid residues, or acyl chains of specifically bound regulatory lipids such as phosphatidylinositol phosphates. Hydrophobic interactions have been shown to be important even for highly cationic peptides and proteins, such as the polybasic domain of the MARCKS protein or histactophilin, when their natural hydrophobic anchors are present.

Covalently Bound Lipid Anchors

Lipid anchored proteins are covalently attached to different fatty acid acyl chains on the cytoplasmic side of the cell membrane via palmitoylation, myristoylation, or prenylation. At the cell surface, on the opposite side of the cell membrane lipid anchored proteins are covalently attached to the lipids glycosylphosphatidylinositol (GPI) and cholesterol. Protein association with membranes through the use of acylated residues is a reversible process, as the acyl chain can be buried in a protein's hydrophobic binding pocket after dissociation from the membrane. This process occurs within the beta-subunits of G-proteins . Perhaps because of this additional need for structural flexibility, lipid anchors are usually bound to the highly flexible segments of proteins tertiary structure that are not well resolved by protein crystallographic studies.

Specific Protein–lipid Binding

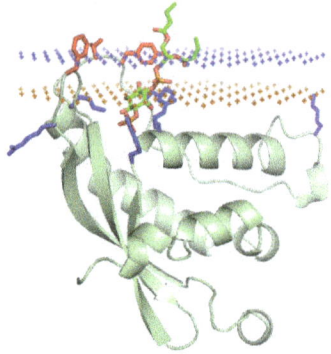

P40phox PX domain of NADPH oxidase Middle plane of the lipid bilayer – black dots. Boundary of the hydrocarbon core region – blue dots (intracellular side). Layer of lipid phosphates – yellow dots.

Some cytosolic proteins are recruited to different cellular membranes by recognizing certain types of lipid found within a given membrane. Binding of a protein to a specific lipid occurs via specific membrane-targeting structural domains that occur within the protein and have specific binding pockets for the lipid head groups of the lipids to which they bind. This is a typical biochemical protein–ligand interaction, and is stabilized by the formation of intermolecular hydrogen bonds, van der Waals interactions, and hydrophobic interactions between the protein and lipid ligand. Such complexes are also stabilized by the formation of ionic bridges between the aspartate or glutamate residues of the protein and lipid phosphates via intervening calcium ions (Ca^{2+}). Such ionic bridges can occur and are stable when ions (such as Ca^{2+}) are already bound to a protein in solution, prior to lipid binding. The formation of ionic bridges is seen in the protein–lipid interaction between both protein C2 type domains and annexins.

Protein–lipid Electrostatic Interactions

Any positively charged protein will be attracted to a negatively charged membrane by nonspecific electrostatic interactions. However, not all peripheral peptides and proteins are cationic, and only certain sides of membrane are negatively charged. These include the cytoplasmic side of plasma membranes, the outer leaflet of outer bacterial membranes and mitochondrial membranes. Therefore, electrostatic interactions play an important role in membrane targeting of electron carriers such as cytochrome c, cationic toxins such as charybdotoxin, and specific membrane-targeting domains such as some PH domains, C1 domains, and C2 domains.

Electrostatic interactions are strongly dependent on the ionic strength of the solution. These interactions are relatively weak at the physiological ionic strength (0.14M NaCl): ~3 to 4 kcal/mol for small cationic proteins, such as cytochrome c, charybdotoxin or hisactophilin.

Spatial Position in Membrane

Orientations and penetration depths of many amphitropic proteins and peptides in membranes are studied using site-directed spin labeling, chemical labeling, measurement of membrane binding affinities of protein mutants, fluorescence spectroscopy, solution or solid-state NMR spectroscopy, ATR FTIR spectroscopy, X-ray or neutron diffraction, and computational methods.

Two distinct membrane-association modes of proteins have been identified. Typical water-soluble proteins have no exposed nonpolar residues or any other hydrophobic anchors. Therefore, they remain completely in aqueous solution and do not penetrate into the lipid bilayer, which would be energetically costly. Such proteins interact with bilayers only electrostatically, for example, ribonuclease and poly-lysine interact with membranes in this mode. However, typical amphitropic proteins have various hydrophobic anchors that penetrate the interfacial region and reach the hydrocarbon interior of the membrane. Such proteins "deform" the lipid bilayer, decreasing the temperature of lipid fluid-gel transition. The binding is usually a strongly exothermic reaction. Association of amphiphilic α-helices with membranes occurs similarly. Intrinsically unstructured or unfolded peptides with nonpolar residues or lipid anchors can also penetrate the interfacial region of the membrane and reach the hydrocarbon core, especially when such peptides are cationic and interact with negatively charged membranes.

Categories

Enzymes

Peripheral enzymes participate in metabolism of different membrane components, such as lipids (phospholipases and cholesterol oxidases), cell wall oligosaccharides (glycosyltransferase and transglycosidases), or proteins (signal peptidase and palmitoyl protein thioesterases). Lipases can also digest lipids that form micelles or nonpolar droplets in water.

Class	Function	Physiology
Alpha/beta hydrolase fold	Catalyzes the hydrolysis of chemical bonds.	Includes bacterial, fungal, gastric and pancreatic lipases, palmitoyl protein thioesterases, cutinase, and cholinesterases
Phospholipase A2 (secretory and cytosolic)	Hydrolysis of sn-2 fatty acid bond of phospholipids.	Lipid digestion, membrane disruption, and lipid signaling.
Phospholipase C	Hydrolyzes PIP2, a phosphatidylinositol, into two second messengers, inositol triphosphate and diacylglycerol.	Lipid signaling
Cholesterol oxidases	Oxidizes and isomerizes cholesterol to cholest-4-en-3-one.	Depletes cellular membranes of cholesterol, used in bacterial pathogenesis.

Carotenoid oxygenase	Cleaves carotenoids.	Carotenoids function in both plants and animals as hormones (includes vitamin A in humans), pigments, flavors, floral scents and defense compounds.
Lipoxygenases	Iron-containing enzymes that catalyze the dioxygenation of polyunsaturated fatty acids.	In animals lipoxygenases are involved in the synthesis of inflammatory mediators known as leukotrienes.
Alpha toxins	Cleave phospholipids in the cell membrane, similar to Phospholipase C.	Bacterial pathogenesis, particularly by *Clostridium perfringens*.
Sphingomyelinase C	A phosphodiesterase, cleaves phosphodiester bonds.	Processing of lipids such as sphingomyelin.
Glycosyltransferases: MurG and Transglycosidases	Catalyzes the transfer of sugar moieties from activated donor molecules to specific acceptor molecules, forming glycosidic bonds.	Biosynthesis of disaccharides, oligosaccharides and polysaccharides (glycoconjugates), MurG is involved in bacterial peptidoglycan biosynthesis.
Ferrochelatase	Converts protoporphyrin IX into heme.	Involved in porphyrin metabolism, protoporphyrins are used to strengthen egg shells.
Myotubularin-related protein family	Lipid phosphatase that dephosphorylates PtdIns3P and PtdIns(3,5)P2.	Required for muscle cell differentiation.
Dihydroorotate dehydrogenases	Oxidation of dihydroorotate (DHO) to orotate.	Biosynthesis of pyrimidine nucleotides in prokaryotic and eukaryotic cells.
Glycolate oxidase	Catalyses the oxidation of α-hydroxy acids to the corresponding α-ketoacids.	In green plants, the enzyme participates in photorespiration. In animals, the enzyme participates in production of oxalate.

Membrane-targeting Domains ("Lipid Clamps")

C1 domain of PKC-delta (1ptr) Middle plane of the lipid bilayer – black dots. Boundary of the hydrocarbon core region – blue dots (cytoplasmic side). Layer of lipid phosphates – yellow dots.

Membrane-targeting domains associate specifically with head groups of their lipid ligands embedded into the membrane. These lipid ligands are present in different concentrations in distinct types of biological membranes (for example, PtdIns3P can be found mostly in membranes of early endosomes, PtdIns(3,5)P2 in late endosomes, and PtdIns4P in the Golgi). Hence, each domain is targeted to a specific membrane.

- C1 domains bind diacylglycerol and phorbol esters.

- C2 domains bind phosphatidylserine or phosphatidylcholine

- Pleckstrin homology domains , PX domains , and Tubby domains bind different phosphoinositides

- FYVE domains are more specific for PtdIns3P.

- ENTH domains bind PtdIns(3,4)P2 or PtdIns(4,5)P2.

- ANTH domain binds PtdIns(4,5)P2.

- Proteins from ERM (ezrin/radixin/moesin) family bind PtdIns(4,5)P2.

- Other phosphoinositide-binding proteins include phosphotyrosine-binding domain and certain PDZ domains. They bind PtdIns(4,5)P2.

- Discoidin domains of blood coagulation factors

- ENTH, VHS and ANTH domains

Structural Domains

Structural domains mediate attachment of other proteins to membranes. Their binding to membranes can be mediated by calcium ions (Ca^{2+}) that form bridges between the acidic protein residues and phosphate groups of lipids, as in annexins or GLA domains.

Class	Function	Physiology
Annexins	Calcium-dependent intracellular membrane/ phospholipid binding.	Functions include vesicle trafficking, membrane fusion and ion channel formation.
Synapsin I	Coats synaptic vesicles and binds to several cytoskeletal elements.	Functions in the regulation of neurotransmitter release.
Synuclein	Unknown cellular function.	Thought to play a role in regulating the stability and/or turnover of the plasma membrane. Associated with both Parkinson's disease and Alzheimer's disease.
GLA-domains of the coagulation system	Gamma-carboxyglutamate (GLA) domains are responsible for the high-affinity binding of calcium ions.	Involved in function of clotting factors in the blood coagulation cascade.
Spectrin and α-actinin-2	Found in several cytoskeletal and microfilament proteins.	Maintenance of plasma membrane integrity and cytoskeletal structure.

Transporters of Small Hydrophobic Molecules

These peripheral proteins function as carriers of non-polar compounds between different types of cell membranes or between membranes and cytosolic protein complexes. The transported substances are phosphatidylinositol, tocopherol, gangliosides, glycolipids, sterol derivatives, retinol, fatty acids, water, macromolecules, red blood cells, phospholipids, and nucleotides.

- Glycolipid transfer proteins

- Lipocalins including retinol binding proteins and fatty acid-binding proteins

- Polyisoprenoid-binding protein

- Ganglioside GM2 activator proteins

- CRAL-TRIO domain (α-Tocopherol and phosphatidylinositol sec14p transfer proteins)

- Sterol carrier proteins

- Phosphatidylinositol transfer proteins and STAR domains

- Oxysterol-binding protein

Electron Carriers

These proteins are involved in electron transport chains. They include cytochrome c, cupredoxins, high potential iron protein, adrenodoxin reductase, some flavoproteins, and others.

Polypeptide Hormones, Toxins, and Antimicrobial Peptides

Many hormones, toxins, inhibitors, or antimicrobial peptides interact specifically with transmembrane protein complexes. They can also accumulate at the lipid bilayer surface, prior to binding their protein targets. Such polypeptide ligands are often positively charged and interact electrostatically with anionic membranes.

Some water-soluble proteins and peptides can also form transmembrane channels. They usually undergo oligomerization, significant conformational changes, and associate with membranes irreversibly. 3D structure of one such transmembrane channel, α-hemolysin, has been determined. In other cases, the experimental structure represents a water-soluble conformation that interacts with the lipid bilayer peripherally, although some of the channel-forming peptides are rather hydrophobic and therefore were studied by NMR spectroscopy in organic solvents or in the presence of micelles.

Class	Proteins	Physiology
Venom toxins	• Scorpion venom • Snake venom • Conotoxins • Poneratoxin (insect)	Well known types of biotoxins include neurotoxins, cytotoxins, hemotoxins and necrotoxins. Biotoxins have two primary functions: predation (snake, scorpion and cone snail toxins) and defense (honeybee and ant toxins).
Sea anemone toxins	• Sea anemone sodium channel inhibitory toxin • Neurotoxin III • Cytolysins	Inhibition of sodium and potassium channels and membrane pore formation are the primary actions of over 40 known Sea anemone peptide toxins. Sea anemone are carnivorous animals and use toxins in predation and defense; anemone toxin is of similar toxicity as the most toxic organophosphate chemical warfare agents.
Bacterial toxins	• Perfringolysin O • Botulinum toxin B • Heat-stable enterotoxin B • δ-Endotoxins • Bacteriocins • Lantibiotic peptides • Gramicidin S	Microbial toxins are the primary virulence factors for a variety of pathogenic bacteria. Some toxins, are Pore forming toxins that lyse cellular membranes. Other toxins inhibit protein synthesis or activate second messenger pathways causing dramatic alterations to signal transduction pathways critical in maintaining a variety of cellular functions. Several bacterial toxins can act directly on the immune system, by acting as superantigens and causing massive T cell proliferation, which overextends the immune system. Botulinum toxin is a neurotoxin that prevents neuro-secretory vesicles from docking/fusing with the nerve synapse plasma membrane, inhibiting neurotransmitter release.
Fungal Toxins	• Cyclic lipopeptide antibiotics Surfactin and daptomycin • Peptaibols	These peptides are characterized by the presence of an unusual amino acid, α-aminoisobutyric acid, and exhibit antibiotic and antifungal properties due to their membrane channel-forming activities.
Antimicrobial peptides	• HP peptide • Saposin B and NK-lysin • Lactoferricin B • Magainin , Moricins , and Pleurocidin	The modes of action by which antimicrobial peptides kill bacteria is varied and includes disrupting membranes, interfering with metabolism, and targeting cytoplasmic components. In contrast to many conventional antibiotics these peptides appear to be bactericidal instead of bacteriostatic.
Defensins	• Insect defensins • Plant defensins : Cyclotides and thionins	Defensins are a type of antimicrobial peptide; and are an important component of virtually all innate host defenses against microbial invasion. Defensins penetrate microbial cell membranes by way of electrical attraction, and form a pore in the membrane allowing efflux, which ultimately leads to the lysis of microorganisms.

Neuronal peptides	• Tachykinin peptides	These proteins excite neurons, evoke behavioral responses, are potent vasodilatators, and are responsible for contraction in many types of smooth muscle.
Apoptosis regulators	• Bcl-2	Members of the Bcl-2 family govern mitochondrial outer membrane permeability. Bcl-2 itself suppresses apoptosis in a variety of cell types including lymphocytes and neuronal cells.

Integral or Intrinsic Membrane Proteins

Integral membrane proteins are associated with membranes and interact strongly with the hydrophobic part of the phospholipid bilayer. Presence of one or more apolar regions accounts for the span of lipid bilayer (α-helix and β-sheet as well). They interact mainly through van der Waals interaction with the hydrophobic core of the lipid bilayer. Thus they can be extracted from the membrane only through membrane disruption by detergents. Examples: GPCRs, rhodposins proteins etc.

Peripheral or Extrinsic Membrane Proteins

Peripheral or extrinsic membrane proteins are known to interact either non covalently with the membrane surface through electrostatic or hydrogen bonds or with covalent bonds through lipids or GPI (glycosylphosphatidylinositol) anchors. They interact with the hydrophilic surfaces of the bilayer through electrostatic interaction. They can be isolated from the membrane using strong salt or change in pH. Examples: Cytochrome C protein.

Type I Membrane Proteins

This is a single-pass transmembrane protein. The N-terminus of this protein is extracellular (luminal) and C-terminus remains in the cytoplasmic region for a cell (or organelle) membrane.

Type II Membrane Proteins

This is a single-pass transmembrane protein. The C-terminus of this protein is extracellular (luminal) and N-terminus remains in the cytoplasmic region for a cell (or organelle) membrane.

Multipass Transmembrane Proteins

Multipass transmembrane proteins are able to cross the lipid bilayer multiple times compared to Type I and Type II single pass membrane proteins which can cross the lipid bilayer only once. Membrane straddling region of polypeptide chains possess mostly α-helical conformation as in the lipid environment hydrogen bonding between polypeptide chains would be maximum if it form helical conformation.

Lipid Chain Anchored-membrane Proteins

Lipid chain anchored-membrane proteins are related with lipid bilayer via one or greater than one covalently attached fatty acid chains or prenyl groups (other type of lipid chains).

GPI-anchored-Membrane Proteins

GPI-anchored-membrane proteins are associated with lipid bilayer via glycosylphosphatidylinositol (GPI) anchor.

Structure of Membrane Proteins

Biological Membranes

Before going into the details of the membrane proteins we need to look at the structural aspect of the biological membranes. Biological membranes were considered to be two dimensional fluids consist of two 'leaflets' which comprised of mainly lipid molecules. According to the fluid mosaic model, the outer part is made up of hydrophilic (ionic and polar head groups) groups which interact with the aqueous solvents. The inner part is comprised of hydrocarbon chains of the lipid. The fluid mosaic model considers membrane as a dynamic system where both proteins and lipids can move and interact.

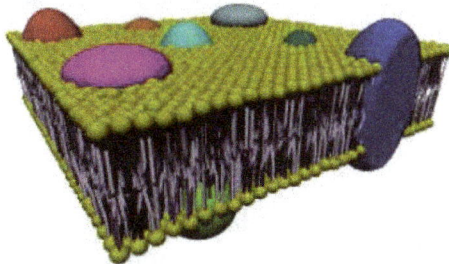

Role of Length and Magnitude of Dielectric Gradient of Membrane Bilayer Thickness

The variation in the dielectric constant between the ionic part and the hydrophobic part (~80 to 2 Debye) is significant which occurs over a comparatively short distance and thus cover up of charge or leaving a hydrogen bond unsatisfied is unfavored. Peptide backbone of a protein is comprised of polar amino and carbonyl groups and thus covering of a peptide backbone in the membrane interior leaving the hydrogen bonding unsatisfied is energetically unfavorable.

Thickness of the bilayer governs the length of low dielectric well which in turn determines interior, exterior and interfacial regions of proteins. Only specific conformations of proteins get stabilized according to the bilayer thickness. There should be comparable length factor between the hydrophobic thickness of the bilayer and the hydrophobic length of membrane protein. It further regulates self aggregation of protein to minimize the unfavorable interactions if the comparable relationship is absent.

Hydrophobic chain packing plays an important role in stabilizing protein structure. The favored arrangement of the hydrophobic chains of bilayer is when they are aligned to each other maximizing van der Waals interactions. Thus, cylindrical shape of the protein will be able to minimize number of lipid chains disordered by their presence and also area of the proteins exposed to the bilayer.

1- The need to match the dielectric properties of the side chains with the lipid.

2- The need to satisfy hydrogen bonds.

3- The need to align with the packing of the hydrocarbon chains.

Structural Classifications: Primary, Secondary, Tertiary and Quaternary Primary Structure

The interior of the membrane is nonpolar in nature. Therefore, surface residues of transmembrane proteins are expected to be nonpolar in nature so that they can reside in the interior part of the membrane. A hydrophobicity scale was formulated to assign numerical values to the hydrophobic nature of amino acid side chains for the prediction purpose. Numerous scale such as Kyte-Doolittle scale, GES scale, etc have been introduced to postulate which part of the proteins will reside in the inner part of the membrane.

Secondary Structure

The inner part of the membrane is devoid of water. Thus the only possibility of the atoms of the peptide backbone to undergo hydrogen bonding is either side chain atoms or other atoms on the peptide backbone. The most favored arrangements are α-helical or β-sheet arrangement as regular arrays of hydrogen bonding occurs between amide nitrogen and carbonyl oxygen atoms. Most of the membrane proteins are found to be helical in nature. This preference for helical structure over β-sheet arrangement might be due to the following reasons:

(1) Helix length is sufficient to accommodate any little changes in bilayer thickness.

(2) Individual insertion of helices is possible whereas before insertion β-strands have to be aligned or zipper up to form sheets.

Tertiary Structure

Folding of integral membrane proteins occurs via a two step process. This picture is clear from the structure of transmembrane helical protein glycophorin. In the first step insertion and formation of helices followed by association of transmembrane helices in the second step. In case β-sheet proteins, first formation of β-sheet occurs followed by insertion in the membrane.

Similar type of packing is observed in the packing of membrane proteins, hydrogen bonding between the helices is less in number and salt bridges are absent.

Mechanism of association of transmembrane helices in the membrane interior Two proposed mechanisms are there:

1. Arrangement of nonpolar side chains of proteins takes place resulting maximum packing of helices

2. Polar and hydrogen bonding side chains of proteins arrange such that they will stabilize interaction between the helices

As rise/residue is ~3.6 for a helix, thus atleast one residue should be polar for interaction between two helices and two or more residues for multiple helix interaction. To explore which side of the helix interact with the membrane interior and which side interacts with other helix a term hydrophobic moment has been introduced by Eisenberg and co-workers. The hydrophobic moment is a vector of the sum of the hydrophobicity of the particular residues on the helix times the unit vector from the nucleus of the α-carbon to the center of the side chain.

$$\vec{\mu}_s = \sum_{n=1}^{n} H_n \vec{S}_n$$

3. Helical moment which arises due to the particular configuration of peptide bonds in helix structure that favors association of helices. Peptide bonds have weak dipole which arises due to their resonance structure. Alignment of these moments within helix results in an overall moment. Antiparallel structure of helix is preferred over parallel structure. More the length of the helix more will be the moment. It has been observed that transmembrane helices have mostly antiparallel configuration.

4. Optimal packing of the helices also governs association of the helices in the bilayer. Angle between membrane spanning proteins and bilayer plane is ~21 ° which is 20 ° for the angle between two membrane spanning proteins. A knob-into-hole packing arrangement of helices results left-handed coiled-coil arrangement of proteins.

Quaternary Structure

Membrane protein oligomerization has been explored by measuring the dimerization of the single transmembrane protein glycophorin with the association of bacteriorhodopsin helices in bilayers. The significance of packing interactions between the helices was recognized. Our knowledge about the thermodynamics, kinetics and several other physical properties that direct the oligomerization of integral membrane proteins is inadequate due to the lack of methods to monitor protein-protein interaction in the membrane.

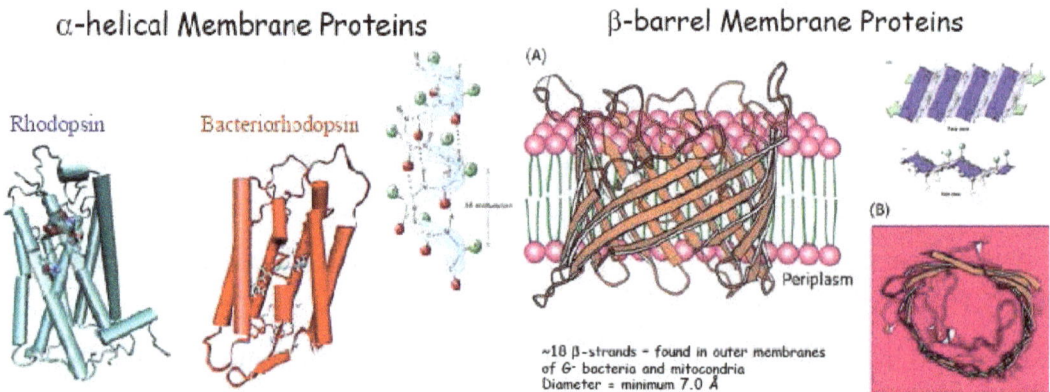

α-helical Membrane Proteins

Rhodopsin   Bacteriorhodopsin

β-barrel Membrane Proteins

(A)

(B)

Periplasm

~18 β-strands - found in outer membranes of G- bacteria and mitocondria
Diameter = minimum 7.0 Å

Membrane Proteins II

In the previous chapter we have discussed the classification and structural aspect of membrane proteins. In this chapter, we have described the transportation across the cell membrane by membrane proteins. Plasma membrane act as a selective permeable membrane which allows necessary molecules to cross in and waste molecules to come out of the cell, often one at a time but also molecules can enter in large packages.

Mechanism of Membrane Protein Transport

Mechanism of transportation can be categorized as follows:

- Active transport: Expense of cellular energy in the form of ATP hydrolysis is responsible for active transport

- Passive transport: Kinetic energy of the molecules being responsible for transportation across the membrane or by transporters

Active Transport Across the Cell Membrane

There is a general tendency of any molecule to move from higher concentration region to lower concentration region due to thermal energy and their motion. This is commonly known as diffusion. Cell membranes behave as impermeable to some molecules due to their size, polarity etc and thus allows only solvent molecules to move across the cell which is known as osmosis and the pressure is known as osmotic pressure.

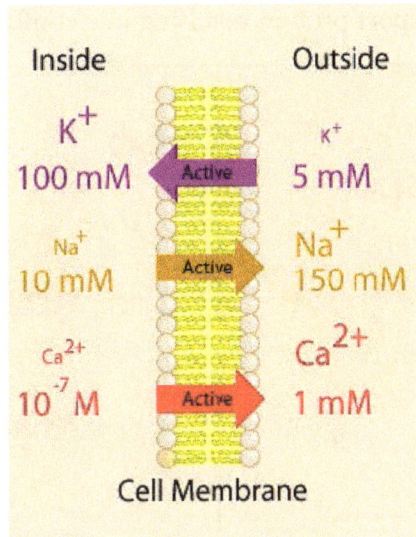

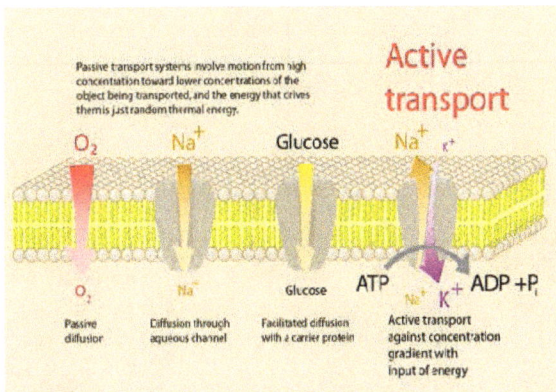

Therefore, it clear that for transportation in reverse manner that is from lower concentration to higher concentration region is possible only when some energy will be spent to overcome the effect of diffusion and osmosis.

For maintenance of the normal state of a cell, large concentration differences in K^+, Na^+ and Ca^{2+} are required and active transport mechanisms comes into account.

The Sodium-Potassium Pump

The sodium and potassium ion pump works across the cell membrane involving the hydrolysis of ATP which provides the necessary energy, is an active transport process. An enzyme known as Na^+/K^+-ATPase is involved in this process. In this process, large excess of Na^+ outside the cell and the large excess of K^+ ions inside the cell are maintained. A schematic diagram of the transport process is given below. Transportation process involves three Na^+ ions to the outside of the cell and two K^+ ions to the inside of the cell. Thus an unbalanced charge transfer occurs and provides charge separation across the membrane. This pump is commonly known as a P-type ion pump. ATP interactions phosphorylate transport protein resulting in a conformational change.

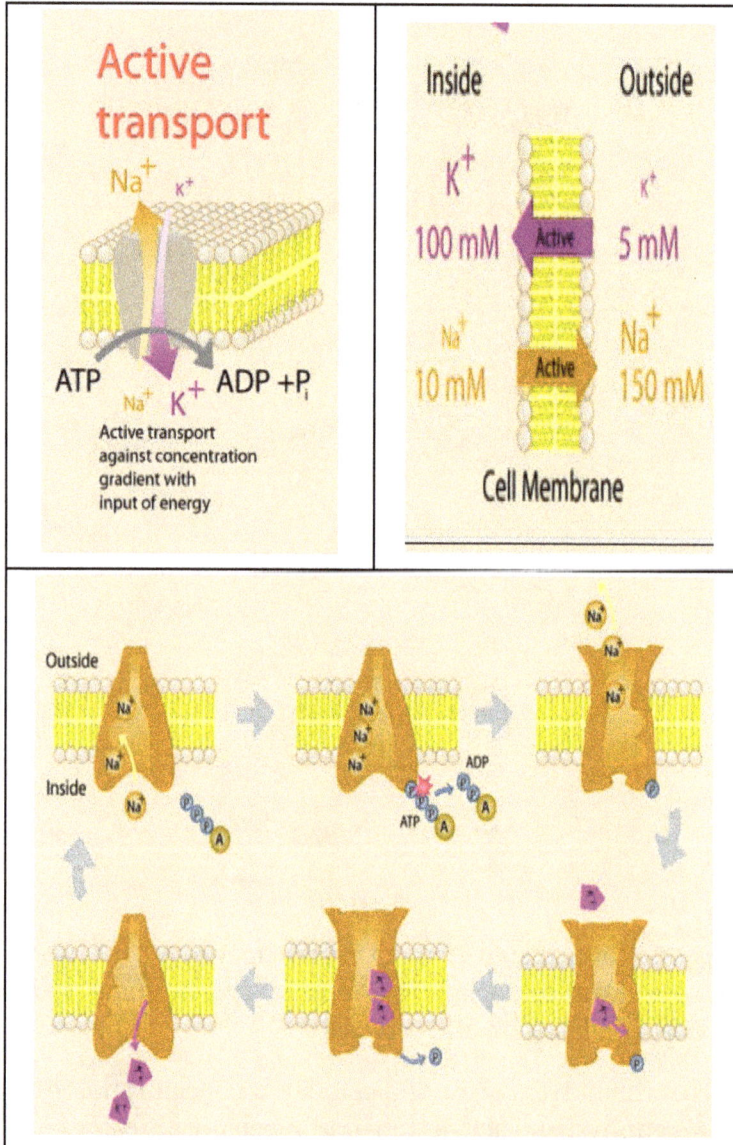

Proton Gradient Across Cell Membrane

In the membrane of both mitochondria and chloroplast, a crucial active transport process is involved in the electron transport process which involves transport of a proton to generate a proton gradient. The energy required for the phosphorylation of ATP by ATP synthase is supplied by this proton gradient.

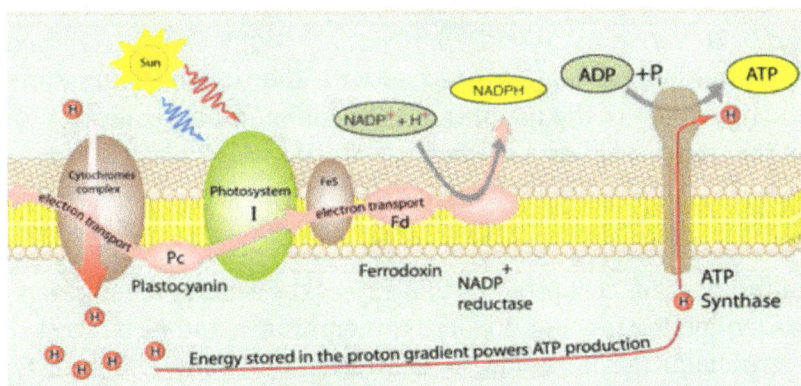

Mitochondrial membrane requires production of ATP as energy currency for carrying out different cell processes. The thylakoid membrane of chloroplast involves electron transport process and transportation of electrons to cytochrome complex which drives protons to move opposite in the direction of the concentration gradient which ultimately provides energy required for ADP to ATP conversion.

Mitochondrial Membrane Transport Proteins

Mitochondrial membrane transport proteins exists in the membrane of mitochondria and responsible for transportation of different molecules or ions across the membrane either inside or outside of the cell.

The mitochondrial permeability transition pore which is known as MPT pore is basically a protein pore. It forms in the inner membrane of mitochondria under certain pathological conditions and in response to enhanced mitochondrial calcium (Ca^{2+}) load and oxidative stress opens up (Crompton, 1999).

The mitochondrial calcium uniporter is known to transport calcium ions from the cytosol of the cell into the mitochondrial matrix (Crompton, 1999; Nicholls, 2005).

The mitochondrial sodium/calcium exchanger in involved in carrying Ca^{2+} ions in exchange for Na^+ ions. These transport proteins play a crucial role to keep a balance between proper electrical and chemical gradients in mitochondria by transporting ions and other factors between the inside and outside of mitochondria.

Glucose Transporters

This consists of a large group of membrane proteins which assist in the transportation of glucose over a plasma membrane. These transporters are present in all phyla as glucose is considered to be a vital source of energy for life. In most of the mammalian cells, GLUT or SLC2A protein family is found. GLUTs are basically integral membrane proteins, consists of 12 membrane- spanning helices along with the exposed amino and carboxyl termini on the cytoplasmic side of the plasma membrane. GLUT proteins transport glucose and related hexoses according to a model of alternate conformation (Oka et al., 1990; Hebert et al., 1992;Cloherty et al., 1995). Glucose binding to one site imparts a conformational change associated with the transportation, and glucose is released to the other side of the membrane. The inner and outer glucose-binding sites are located in the transmembrane segments 9, 10, 11.

Neurotransmitter Transporter

A class of membrane transport proteins that extent the cellular membranes of neurons. They mainly carry neurotransmitters across membranes and also direct their further transport to specific intracellular locations. More than twenty different types of neurotransmitter transporters are known to exist (Iversen, 2000).

Vesicular transporters shift neurotransmitters into synaptic vesicles and also regulate the concentrations of substances (Johnson, 2003). They depend on proton gradient which originates by the hydrolysis of adenosine triphosphate (ATP) to carry out their work such as vesicle ATPase hydrolyzes ATP, enables protons to get pumped into the Synaptic vesicles and creates a proton gradient. Thus the efflux of protons from the vesicle supplies the required energy to transport the neurotransmitter into the vesicle (Kandel, 2000).

Neurotransmitter can be Classified as Follows:

- Glutamate/aspartate transporters

- GABA transporters

- Glycine transporters,

- Monoamine transporters

- Adenosine transporters

- Vesicular acetylcholine transporter

ATP-binding Cassette Transporters (ABC-transporter)

ABC-transporters are the members of a protein superfamily which is one of the largest and most ancient families (Jones et al., 2004; Ponte-Sucre, 2009). They are transmembrane proteins; exploit the energy obtained from hydrolysis of adenosine triphosphate (ATP) to perform various *in vivo* processes and also translocation of various substrates across cell membranes including non-transport-related processes such as translation of RNA and DNA repair. They are known to involve in the transportation of a large variety of substrates across extra-and intracellular membranes, including metabolic products, lipids and sterols, and drugs.

Ion Transporter

An ion transporter, commonly known as an ion pump, is a transmembrane protein that regulates movement of ions across a plasma membrane against their concentration gradient which is in contrast to the ion channels, where ions move through passive transport. Energy from different sources including ATP, sunlight, and other redox reactions is converted into potential energy by these primary transporters (enzymes). Secondary transporters use this energy, including ion carriers and ion channels, to regulate vital cellular processes, such as ATP synthesis.

Example of a particular kind of ion pump is Na^+/K^+-ATPase.

Facilitated Diffusion (Passive-mediated Transport)

It is a process of passive transport (as opposed to active transport) which is supported by integral membrane proteins which involves spontaneous passage of molecules or ions across a biological membrane. The diffusion process occurs either across the biological membranes or through the aqueous compartments of an organism (Pratt, 2002). Polar molecules and charged ions exist in water in the dissolved condition although they are unable to diffuse freely across the plasma membrane due to the hydrophobic nature of the fatty acid tails of phospholipids lipid bilayers. Polar molecules are transported across the membranes by proteins which are known to form transmembrane channels which are gated and can regulate the flow of ions or small polar molecules. Poorly water soluble polar molecules such as retinol either through aqueous compartments of cells or through extracellular space by water-soluble carriers as retinol binding proteins.

Donnan Equilibrium: Membrane Transport

Definition

The Gibbs-Donnan equilibrium is defined as uneven distribution of charges on one side of a semi-permeable membrane. Charged particles sometimes failed to pass through the membrane and thus create uneven charge distribution which is the origin of this equilibrium.

It helps in the understanding of the functions of living cells.

How Does this Equilibrium Work?

When two solutions of different concentration are separated by semi-permeable membrane diffusion occurs from high concentration to low concentration region which equalizes their concentration on both sides of the membrane. However, when impermeable solute is present in any one of the solution, then even at the equilibrium with respect to the impermeable solute solution concentration remains high and no equalization is achieved. This type of equilibrium is called 'Donnan equilibrium'.

Donnan Equilibrium: *In Vivo*

Cell membranes are very selective and thus they allow only specific substances to pass through the membrane. Donnan equilibrium directs flow of different6 solutes and ions across the cell membrane and its environment. In the living cell, impermeable solutes (anionic colloids) exist which are mostly made up of proteins and organic phosphate. As they can not cross the cell membranes, concentration of non-diffusible ions across the membranes is high and thus Donnan equilibrium is created across the cell membrane. As a result of this, concentration of ions is more inside the cells.

How Does Cell Maintains this Differential Ion Concentration?

Sodium pump (Na^+-K^+ ATPase) in the cell membrane maintains ion distribution across the cell and prevents cell from disruption via continuously removing excess ions. As the membrane is low permeable to sodium, the pump along with the membrane resists sodium from entering into the cell. The sodium pump enables the membrane impermeable to sodium, which establishes Donnan Equilibrium. The Donnan effect created due to the impermeable anionic colloids maintains the balance with the Donnan effect originated due to the impermeable extracellular sodium.

Gibbs–Donnan Effect

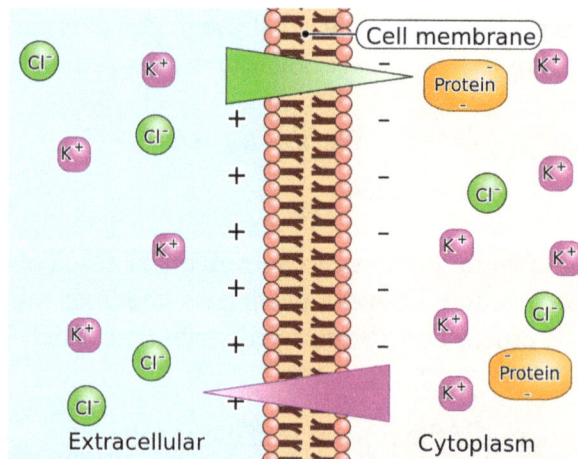

Donnan equilibrium across a cell membrane (schematic)

The Gibbs–Donnan effect (also known as the Donnan's effect, Donnan law, Donnan equilibrium, or Gibbs–Donnan equilibrium) is a name for the behaviour of charged particles near a semi-permeable membrane that sometimes fail to distribute evenly across the two sides of the membrane. The usual cause is the presence of a different charged substance that is unable to pass through the membrane and thus creates an uneven electrical charge. For example, the large anionic proteins in blood plasma are not permeable to capillary walls. Because small cations are attracted, but are not bound to the proteins, small anions will cross capillary walls away from the anionic proteins more readily than small cations.

Some ionic species can pass through the barrier while others cannot. The solutions may be gels or colloids as well as solutions of electrolytes, and as such the phase boundary between gels, or a gel and a liquid, can also act as a selective barrier. The electric potential arising between two such solutions is called the Donnan potential.

The effect is named after the American physicist Josiah Willard Gibbs and the British chemist Frederick G. Donnan.

The Donnan equilibrium is prominent in the triphasic model for articular cartilage proposed by Mow and Lai, as well as in electrochemical fuel cells and dialysis.

The Donnan effect is extra osmotic pressure attributable to cations (Na^+ and K^+) attached to dissolved plasma proteins.

Example

The presence of a charged impermeant ion (for example, a protein) on one side of a membrane will result in an asymmetric distribution of permeant charged ions. The Gibbs–Donnan equation at equilibrium states (assuming permeant ions are Na^+ and Cl^-):

$$\left[Na_{Side\ 1} \right] \times \left[Cl_{Side\ 1} \right] = \left[Na_{Side\ 2} \right] \times \left[Cl_{Side\ 2} \right]$$

Example--

Start	Equilibrium	Osmolarity
Side 1: 9 Na, 9 Cl	Side 1: 6 Na, 6 Cl	Side 1: 12
Side 2: 9 Na, 9 Protein	Side 2: 12 Na, 3 Cl, 9 Protein	Side 2: 24

Double Donnan

Note that Sides 1 and 2 are no longer in osmotic equilibrium (i.e. the total osmolytes on each side are not the same)

In vivo, ion balance does not equilibriate at the proportions that would be predicted by the Gibbs-Donnan model, because the cell cannot tolerate the attendant large influx of water. This is balanced by instating a functionally impermeant cation extracellularly to counter the anionic protein, Na^+. Na^+ does cross the membrane via leak channels (the permeability is approximately 1/10 that of K^+, the most permeant ion) but, as per the pump-leak model, it is extruded by the Na^+/K^+-ATPase.

pH Change

Because there is a difference in concentration of ions on either side of the membrane, the pH may also differ when protons are involved. In many instances, from ultrafiltration of proteins to ion exchange chromatography, the pH of the buffer adjacent to the charged groups of the membrane is different from the pH of the rest of the buffer solution. When the charged groups are negative (basic), then they will attract protons so that the pH will be lower than the surrounding buffer. When the charged groups are positive (acidic), then they will repel protons so that the pH will be higher than the surrounding buffer.

Physiological Applications

Red Blood Cells

When tissue cells are in a protein-containing fluid, the Donnan effect of the cytoplasmic proteins is equal and opposite to the Donnan effect of the extracellular proteins. The opposing Donnan effects cause chloride ions to migrate inside the cell, increasing the intracellular chloride concentration. The Donnan effect may explain why some red blood cells do not have active sodium pumps; the effect relieves the osmotic pressure of plasma proteins, which is why sodium pumping is less important for maintaining the cell volume .

Neurology

Brain tissue swelling, known as cerebral oedema, results from brain injury and other traumatic head injuries that can increase intracranial pressure (ICP). Negatively charged molecules within cells create a fixed charge density, which increases intracranial pressure through the Donnan effect. ATP pumps maintain a negative membrane potential even though negative charges leak across the membrane; this action establishes a chemical and electrical gradient.

The negative charge in the cell and ions outside the cell creates a thermodynamic potential; if damage occurs to the brain and cells lose their membrane integrity, ions will rush into the cell to balance chemical and electrical gradients that were previously established. The membrane voltage will become zero, but the chemical gradient will still exist. To neutralize the negative charges within the cell, cations flow in, which increases the osmotic pressure inside relative to the outside of the cell. The increased osmotic pressure forces water to flow into the cell and tissue swelling occurs.

References

- Harvey, SC (2015). "The scrunchworm hypothesis: Transitions between A-DNA and B-DNA provide the driving force for genome packaging in double-stranded DNA bacteriophages". Journal of Structural Biology. 189: 1–8. doi:10.1016/j.jsb.2014.11.012. PMC 4357361. PMID 25486612

- Shepherd, V. A. (2006). "The cytomatrix as a cooperative system of macromolecular and water networks". Current Topics in Developmental Biology. Current Topics in Developmental Biology. 75: 171–223. doi:10.1016/S0070-2153(06)75006-2. ISBN 9780121531751. PMID 16984813

- Gardiner, Mary Beth (September 2005). "The Importance of Being Cilia" (PDF). HHMI Bulletin. Howard Hughes Medical Institute. 18 (2). Retrieved 2010-03-18

- Holcman, David; Korenbrot, Juan I. (2004). "Longitudinal Diffusion in Retinal Rod and Cone Outer Segment Cytoplasm: The Consequence of Cell Structure". Biophysical Journal. 86 (4): 2566–2582. Bibcode:2004BpJ....86.2566H. doi:10.1016/S0006-3495(04)74312-X. PMC 1304104. PMID 15041693

- Cowan AE, Moraru II, Schaff JC, Slepchenko BM, Loew LM (2012). "Spatial Modeling of Cell Signaling Networks". Methods in Cell Biology. Methods in Cell Biology. 110: 195–221. doi:10.1016/B978-0-12-388403-9.00008-4. ISBN 9780123884039. PMC 3519356. PMID 22482950

- Cho, W. & Stahelin, R.V. (June 2005). "Membrane-protein interactions in cell signaling and membrane trafficking". Annual Review of Biophysics and Biomolecular Structure. 34: 119–151. PMID 15869386. doi:10.1146/annurev.biophys.33.110502.133337. Retrieved 2007-01-23

- Taylor, C. V. (1923). "The contractile vacuole in Euplotes: An example of the sol-gel reversibility of cytoplasm". Journal of Experimental Zoology. 37 (3): 259–289. doi:10.1002/jez.1400370302

- Steven R. Goodman (2008). Medical cell biology. Academic Press. pp. 37–. ISBN 978-0-12-370458-0. Retrieved 24 November 2010

- Lomize A, Lomize M, Pogozheva I. "Comparison with experimental data". Orientations of Proteins in Membranes. University of Michigan. Retrieved 2007-02-08

- Feneberg, Wolfgang; Sackmann, Erich; Westphal, Monika (2001). "Dictyostelium cells' cytoplasm as an active viscoplastic body". European Biophysics Journal. 30 (4): 284–94. doi:10.1007/s002490100135. PMID 11548131

- David S. Cafiso Structure and interactions of C2 domains at membrane surfaces. In: Tamm LK (Editor) (2005). Protein-Lipid Interactions: From Membrane Domains to Cellular Networks. Chichester: John Wiley & Sons. pp. 403–22. ISBN 3-527-31151-3. CS1 maint: Extra text: authors list (link)

- Porter, M.; Sale, W. (2000). "The 9 + 2 axoneme anchors multiple inner arm dyneins and a network of kinases and phosphatases that control motility". The Journal of Cell Biology. 151 (5): F37–F42. doi:10.1083/jcb.151.5.F37. PMC 2174360. PMID 11086017

Functions of Electron and Enzymes

Electron transfer helps in the rearrangement of atoms from one chemical entity to another. Several processes, such as photosynthesis and respiration, which are very important to an organism involve electron transfer. The three main types of electron transfer are heterogeneous electron transfer, inner-sphere electron transfer and outer-space electron transfer. The diverse functions of electrons and enzymes have been thoroughly discussed in this chapter.

Electron Transfer

The sun is main source of energy on the earth. The sun light is consumed by the plant and cyanobacteria via photosynthesis process. In this process CO_2 is fixed to carbohydrate. After that via oxidation process the carbohydrate are metabolized in presence of O_2. These two processes are main reason for life on earth. Or in one sentence one can say these oxidation and reduction processes are the basic primary metabolic reaction step in life. And all these processes are electron transfer process in protein.

Photosynthesis

The photosynthesis is the light driven process by which CO_2 is fixed to produce carbohydrates.

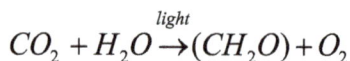

$$CO_2 + H_2O \xrightarrow{light} (CH_2O) + O_2$$

In this process both CO_2 and water are reduced to carbohydrate and oxygen. Photo synthetically produced carbohydrate is the main source of energy for the photosynthetic cell and normal cell. The final ingredients of overall photosynthetic recipe were demonstrated by the German physiologist Robert Mayer who concluded that plants convert light (solar energy) to carbohydrate (chemical energy) from CO_2.

Chloroplasts

The site of the photosynthesis in the eukaryotes (algae and plant) is chloroplast. In chloroplast, light harvesting and carbon assimilation reactions are take place in side the chloroplast. These chloroplasts are surrounded by two membranes, the outer membranes are permeable to small molecule and ions, and an inner membrane which is encloses the internal compartment. This compartment contains many flattened vesicles or sacs known as thylakoids and the aqueous phase enclosed by the inner membrane called as stroma. These thylakoids arranged in stacks called grana. The photosynthetic pigments and enzyme complex are present inside the thylakoid membrane. In the stroma, lots of enzymes are present.

Photosynthesis occurs in two distinct phases:

- The light reactions, which use light energy to generate NADPH and ATP in thylakoid membrane.

- The dark reactions, actually light-independent reactions, which use NADPH and ATP to synthesis carbohydrate from CO_2 and H_2O in stroma.

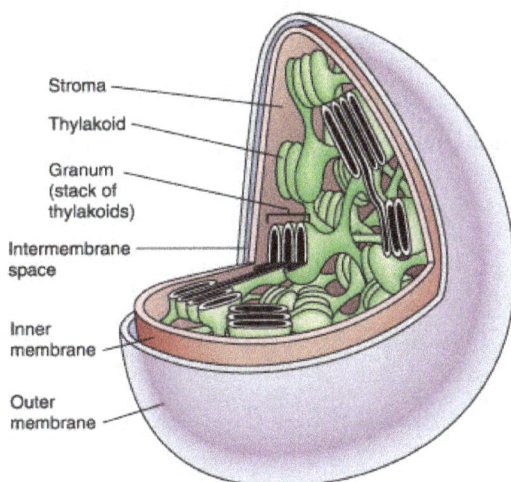

Stroma
Thylakoid
Granum
(stack of
thylakoids)
Intermembrane
space
Inner
membrane
Outer
membrane

Light Reactions

In the first decades of 20th century it was assumed that light was absorbed by the photosynthetic pigments which directly reduced CO_2 to carbohydrate combined with water. In this view in 1931, Corneils van Neil performed photosynthesis process anaerobically using green photosynthetic bacteria in presence of water using H_2S, generate sulfur.

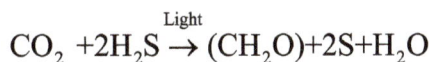

$$CO_2 + 2H_2S \xrightarrow{\text{Light}} (CH_2O) + 2S + H_2O$$

Between the chemical similarity between H_2S and H_2O Neil proposed this general photosynthetic reaction.

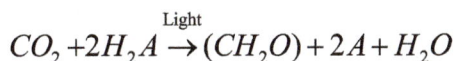

$$CO_2 + 2H_2A \xrightarrow{\text{Light}} (CH_2O) + 2A + H_2O$$

Here H_2A is H_2O in green plants and H_2S in photosynthetic sulfur bacteria. On the basis of this result Neil hypothesized that photosynthesis is the two step process in which light energy is used to dissociate H_2A (light reaction):

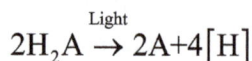

$$2H_2A \xrightarrow{\text{Light}} 2A + 4[H]$$

And the resulting reducing agent [H] subsequently reduced CO_2 to CH_2O and H_2O (the dark reactions):

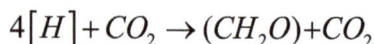

$$4[H] + CO_2 \rightarrow (CH_2O) + CO_2$$

In 1937 Robert Hill found that when leaf extracts containing chloroplasts in presence of non biological electron acceptors like dichlorophenolindophenol or ferricyanide are reduced and oxygen evolved in presence of light. But in the dark neither these reagents are reduced nor oxygen evolved. This was the first evidence that absorbed light energy causes electrons to flow from H_2O to an electron acceptor.

In 1941, when the oxygen isotope became available, Samuel Ruben and Martin Kamen directly demonstrated that the source of the O_2 formed in photosynthesis is H_2O:

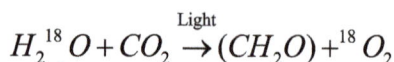

$$H_2^{18}O + CO_2 \xrightarrow{\text{Light}} (CH_2O) + {}^{18}O_2$$

Several years later Severo Ochao showed that $NADP^+$ is the biological electron acceptor in thylakoid membrane of chloroplasts in (light reaction) according to the equation:

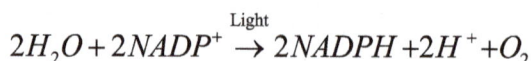

$$2H_2O + 2NADP^+ \xrightarrow{\text{Light}} 2NADPH + 2H^+ + O_2$$

Light Absorption

Visible light is the electromagnetic radiation of wavelengths from 400-700 nm ranging from violet to red, with the former at higher energy and red of lesser energy. The energy of the photon (a quantum of light) follows the Plank equation:

$$E = h\nu = \frac{hc}{\lambda}$$

Where h is the Plank constant (6.626×10^{-34} J.s), ν is the frequency of the light, c is the spped of the light (3×10^8 m/s) λ is the wavelength.

When a photon is absorbed, an electron in the absorbing molecule is lifted to a higher energy level. The energy of absorbed photon (a quantum) exactly matches with the energy of the electronic transition. A molecule that has absorbed a photon is in an excited state, which is generally unstable. An electron lifted to the higher energy orbital level usually returns to its lower energy level via various processes. The excited molecule decays to the stable ground state giving up the absorbed quantum as light or heat or using it to do chemical work. The electron can jump from ground state (S_0) to first (S1), second (S_2) singlet excited state. Also this radiation process is called as fluorescence process. Electron moves from S_2 to S_1 via irradiative path way which is known as internal conversion. Also from S_1 state to electron can move to triplet state (T_1) state irradiative pathway and this process is called internal conversion. From this triplet state to electron can go to the ground state via radiative pathway which known as phosphorescence. Exciton transfer (resonance energy transfer) in which an excited molecule directly transfers its excitation energy to nearby unexcited molecules with similar electronic properties. This process occurs through interactions between the molecular orbitals of the participating molecules in a manner analogous to the interactions between the two pendulums of the similar frequencies.

The amount of light is absorbed by a substance at a given wavelength is described by Beer-Lambert law:

$$A = \log\frac{I_0}{I} = \varepsilon cl$$

Where A is the absorbance, I_0 and I are the intensities of the incident and transmitted light, c is the molar concentration of the sample, l is the length of the light path through the sample in cm and ε is the molar excitation coefficient. Consequently A versus λ plot for a given molecule is called its absorption spectrum.

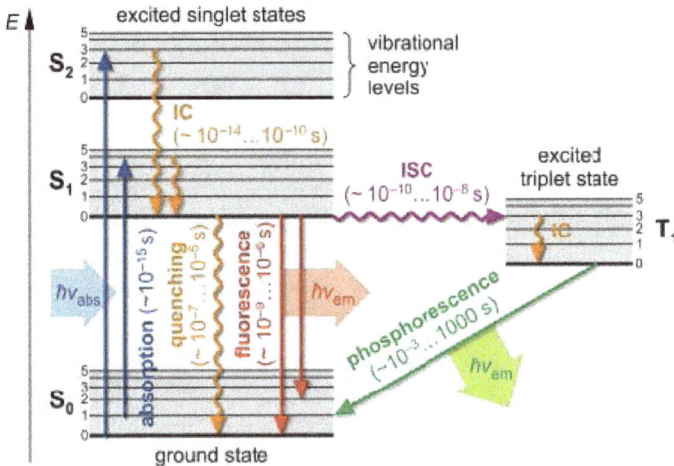

Chlorophylls are the most important light absorbing pigments in the thylakoid membranes. These green pigments are planar, polycyclic containing porphyrin ring containing Mg^{2+}.the heterocyclic five ring system that surrounds the Mg^{2+} has an extended polyene structure with an alternating single and double bonds. These moieties have characteristically high absorption in the visible region of the spectrum. The chlorophylls have high molar extinction coefficient and therefore are well suited for absorbing visible light during photosynthesis.

Chloroplasts contain both chlorophyll a and chlorophyll b. Although both are green in color they are absorbed light at different wavelength complement each other. Normally chlorophylls a are twice with respect to chlorophylls b. The pigments in the algae and cyanobacteria are different slightly from the plant pigment.

These chlorophyll moieties are bound with the protein and formed light-harvesting complexes. The pigments are fixed in relation to each other in other protein complexex and to the membrane. In cyanobacteria and red algae have phycoerythrobilin and phy-cocyanobiln as light harvesting agent. In addition there are light harvesting pigments are presents called carotenoids which are yellow, red and purple in color.

The light harvesting complexes in the thylakoid or bacterial membranes are arranged in a pattern called photo systems. In chloroplasts, each photosystem contains about 200 chlorophyll and 50 carotenoid molecules. All the pigments are able to absorb light but only a few chlorophyll molecules attached with the reaction centre are engaged to transform light energy to chemical energy. The other pigment molecules are called light harvesting or antenna molecules. They absorb light and transmit rapidly and efficiently to the reaction center. In this process a positive charge is formed in one center and in the other center a negative charge is created that forms a potential gap.

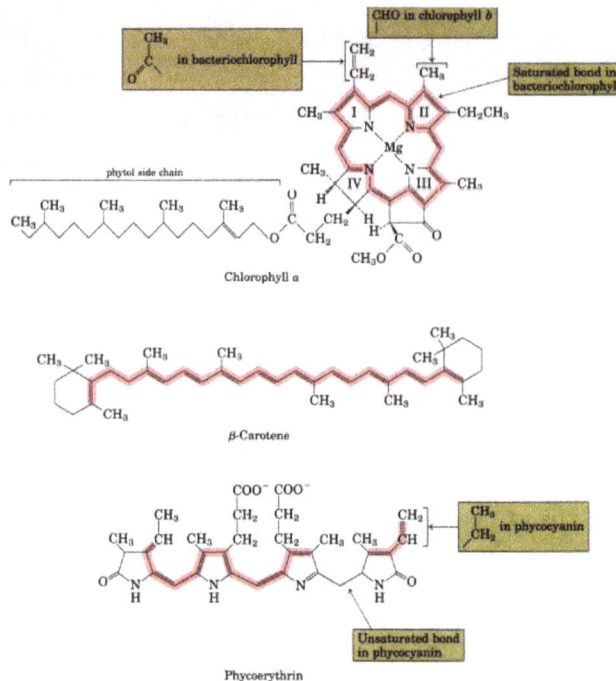

Chlorophyll a

β-Carotene

Phycoerythrin

Electron Transport in Chloroplast

The electron transport in the chloroplast is very complex process. Three alkaloid membrane bound proteins (1) PSII (2) cytochrome b6f complex and (3) PSI are engaged. in this process. The electrons are transferred via mobile electron carriers. The plastoquinone are reduced to plastoquinol in PSII and linked with cytochrome b6f complex. This cytochrome b6f complex is linked with PSI via mobile protein plastocyanin. The electron from PSI is used to reduce $NADP^+$ to NADPH in stroma. And the electron transfer is happened from water to electron whole of P680 and generate O_2 and proton.

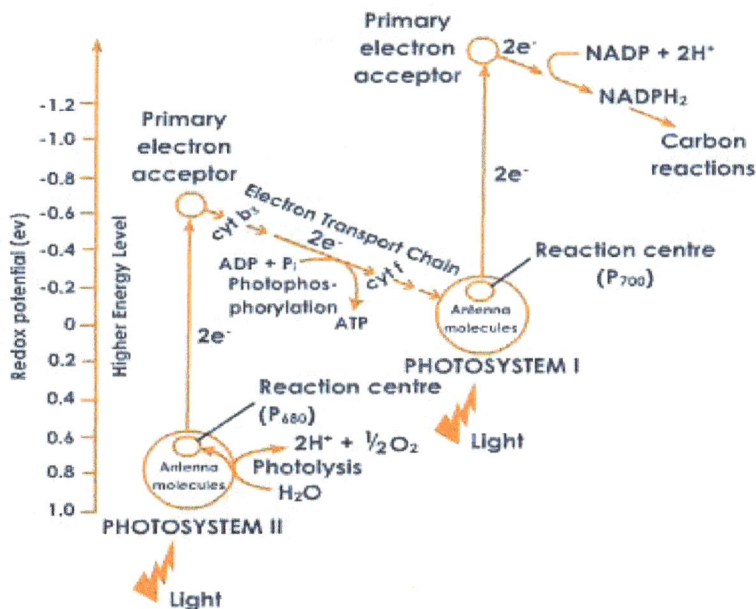

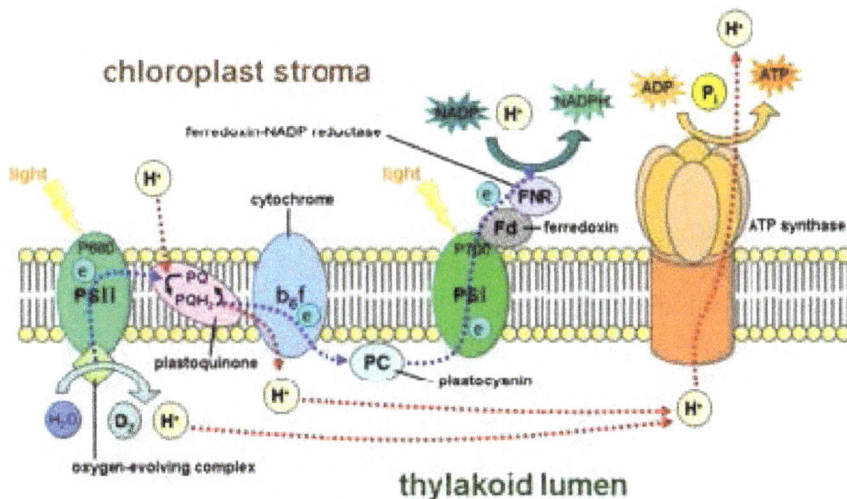

Electron Transport and Oxidative Photophosphorylation:

In 1948, Kennedy and Lehninger discovered that mitochondria are the site of oxidative photophosphorylation. Mitochondria like gram negative bacteria have two membranes, an outer and an inner membrane. The two membranes have different properties. Because of this double-membrane organization, there are five distinct parts to a mitochondrion. They are:

- the outer mitochondrial membrane,

- the intermembrane space (the space between the outer and inner membranes),

- the inner mitochondrial membrane,

- the cristae space (formed by infoldings of the inner membrane), and

- the matrix (space within the inner membrane).

The outer mitochondrial membrane encloses the entire organelle It contains large numbers of integral proteins called porins. These porins form channels that allow molecules 5000 Daltons or less in molecular weight to freely diffuse from one side of the membrane to the other.

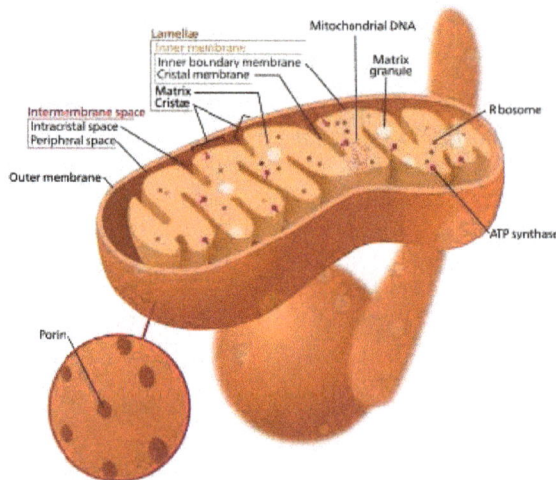

The intermembrane space is the space between the outer membrane and the inner membrane. The inner membrane contains 151 different polypeptides and has a very high protein-to-phospholipid ratio (more than 3:1 by weight). Unlike the outer membrane, the inner membrane doesn't contain porins, and is highly impermeable to all molecules. Almost all ions and molecules require special membrane transporters to enter or exit the matrix. The inner mitochondrial membrane is compartmentalized into many cristase, which expand the surface area of the inner mitochondrial membrane, enhancing its capability to produce ATP. The matrix is the space enclosed by the inner membrane. It contains about 2/3 of the total protein in a mitochondrion. The matrix is important in the production of ATP with the aid of ATP synthase contained in the inner membrane.

Thermodynamics of Electron Transport

We can determine the thermodynamic efficiency of the electron migration based on the knowledge of reduction potential. The thermodynamics of an oxidation-reduction reaction can be measured using the half cell emf value of the reduction process and the oxidation processes in the biological system relative to the standard hydrogen electrode. The standard reduction potential difference (ΔE^0) of a redox reaction was determined from two half cell using the following formula.

$$\Delta E^0 = E^o_{red} + E^0_{ox}$$

Where E^0_{red} is reduction potential of the half cell of reduction process and E^0_{ox} is the oxidation potential of the half cell for oxidation process.

In the biological system NADH is oxidized to NAD in presence of O_2.the overall reaction is

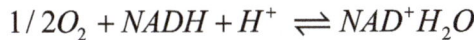

$$1/2O_2 + NADH + H^+ \rightleftharpoons NAD^+ H_2O$$

The half cell for the oxidation process

$$NADH \rightleftharpoons NAD^+ + H^+ + 2e^-$$

$$\Delta E^0 = 0.315V$$

And for the reduction process

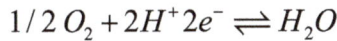

$$1/2\,O_2 + 2H^+ 2e^- \rightleftharpoons H_2O$$

$$\Delta E^0 = 0.815V$$

The toatal emf value of the cell raction is ($\Delta E = 0.815 + .0.315 = 1.130V$) 1.130 volt. The standard free energy for this reaction is calculated from the following equation

$$\Delta G^0 = -nFE^0$$

where n is the no electron transferred to reactants, F is the faraday constant (96494 C·mol⁻¹). So the standard free energy for the above redox reaction is -218 kJ/mol. The standard free energy required to synthesize ATP from ADP and phosphate is 30.5 kJ/mol. Though this process is energetically unfavorable, when this process is coupled with the NADH oxidation process then the overall reaction is energetically favorable. Oxidation of one NADH therefore results in the synthesis of three ATPs.

The Sequence of Electron Transport in Mitochondria

The free energy for ATP synthesis is extracted from the oxidation of NADH and $FADH_2$ by electron. This process is happened in presence of four protein complexes through which electron passes from lower to higher standard reduction potentials. Electrons are transport from complexes I and II to complex III by conenzyme Q (CoQ or ubiquinone) and from Complex III to Complex IV by the peripheral membrane protein cytochrome c.

Complex I catalyzes oxidation of NADH by CoQ:

$$NADH + CoQ \ (oxidised) \longrightarrow NAD^+ + CoQ \ (reduced)$$

$$\Delta E^0 = 0.360V$$

Complex III catalyzes oxidation of CoQ (reduced) by cytochrome c

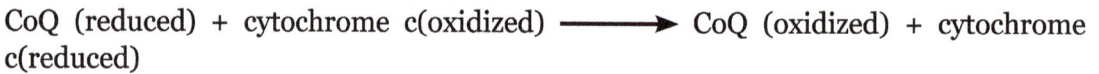

$$CoQ \ (reduced) + cytochrome \ c(oxidized) \longrightarrow CoQ \ (oxidized) + cytochrome \ c(reduced)$$

$$\Delta E^0 = 0.190V$$

Complex IV catalyzes the oxidation of cytochrome c (reduced) by O_2, the terminal electron acceptor of the electron transport chain.

$$cytochrome \ c \ (reduced) + 1/2 \ O_2 \longrightarrow cytochrome \ c \ (oxidized) + H_2O$$

$$\Delta E^0 = 0.580V$$

In this electron transport process in each step the oxidation process is sufficient to power the synthesis of an ATP molecule.

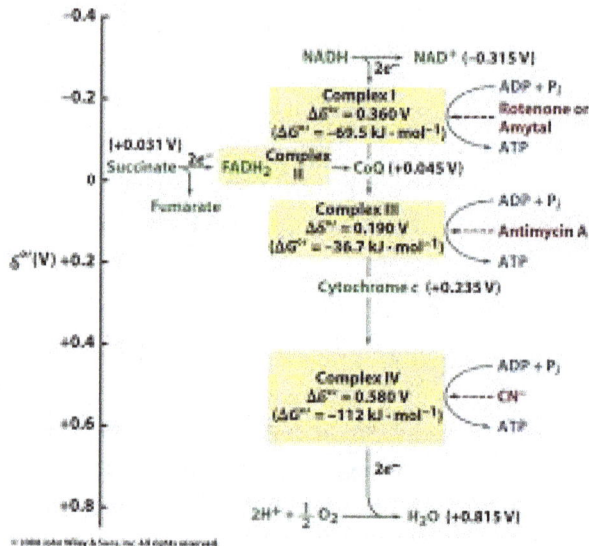

Components of the Electron-Transport Chain

Many of the proteins in the inner mitochondrial membrane are organized into the four respiratory complexes of the electron transport chain. Each cpomplex consist of many protein molecule and contains a vriety of redox active prosthetic group with successively increasing reduction potentials. The complexes are mobile in the mitochondrial membrane. They are not formed a stable structures.

Complex I (NADH-Coenzyme Q Reductase)

Complex I passes electron from NADH to CoQ. It contains one molecule of flavin mononucleotide (FMN a redox active prosthetic group) and six to seven iron-sulfur clusters. These iron sulfur clusters are redox active. There are three types of iron sulfur clusters in the iron-sulfur proteins. The oxidized and reduced state of all iron-sulfur clusters differ by one formal charge regardless of their number of Fe atoms.

The Coenzymes of Complex I

FMN and CoQ are the coenzyme of complex I. These coenzymes have three oxidation states. Though NADH takes participation in two electron transfer system, but these coenzymes can take participate in the one or two electron transfer system because their semiquinone structures are stable. In contrast, the cytochrrome of complex III can only conduct one electron. Therefore FMN and CoQ conduct electron transfer from two electron donor NADH to the one electron acceptor cytochromes.

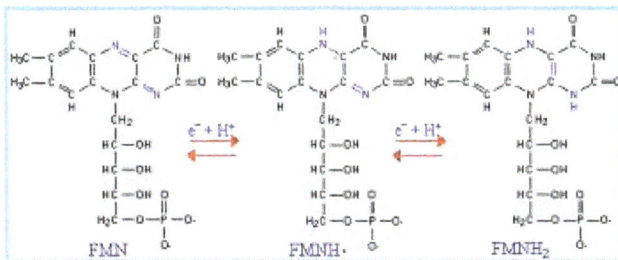

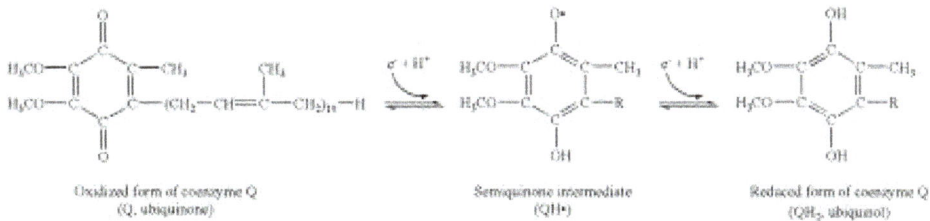

Ubiquinone (Q) is reduced to ubiquinol (QH₂) through a semiquinone intermediate (QH•)

Complex II (Succinate –coenzyme Q Reductase)

Complex II contains the dimeric citric acid cycle enzyme succinate dehydrogenase and three other small hydrophobic subunits. It passes electrons from from succinate to CoQ.

Complex III (Coenzyme Q-Cytochrome c Reductase)

Complex III passes electrons from reduced CoQ to cytochrome c. It contains two b- cytochromes. The number of subunits can be small, as small as three polypeptide chains. This number does increase, and eleven subunits are found in higher animals. Three subunits have prosthetic groups. The cytochrome b subunit has two b-type hemes (bL and bH), the cytochrome c subunit has one c-type heme (c1), and the Rieske Iron Sulfur Protein subunit (ISP) has a two iron, two sulfur iron-sulfur cluster (2Fe•2S).

Cytochromes

Cytochromes are redox active proteins that occur in all organisms. These proteins contain heme groups that reversibly alternate between Fe(II) and Fe(III) oxidation states during electron transport.

The reduced heme groups groups of cytochrome c have three peaks:

α, β, γ bands. The

wavelength of the α peak it is absent in the oxidized species. According the spetra of the mitochondrial membranes they contains three kinds of cytochrome species, cytochrome a, b and c.

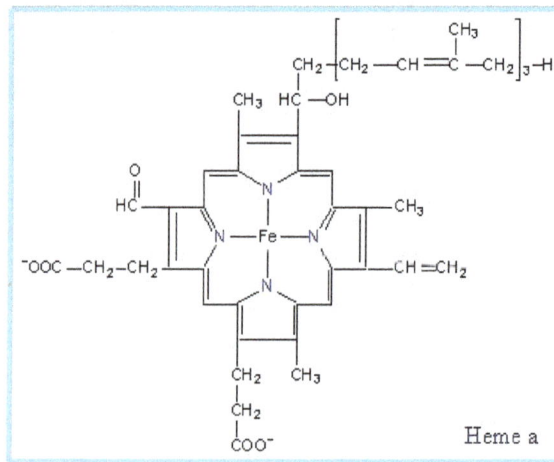

Heme a

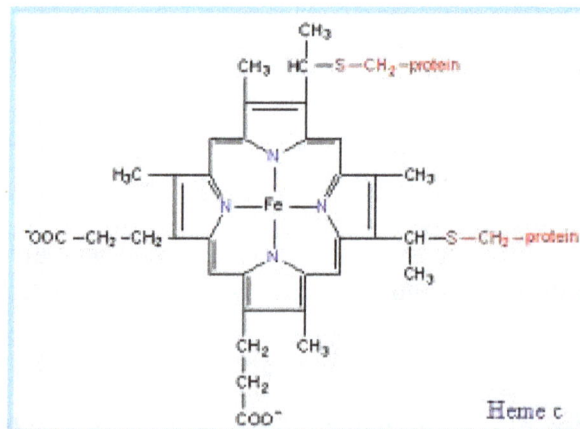

Heme c

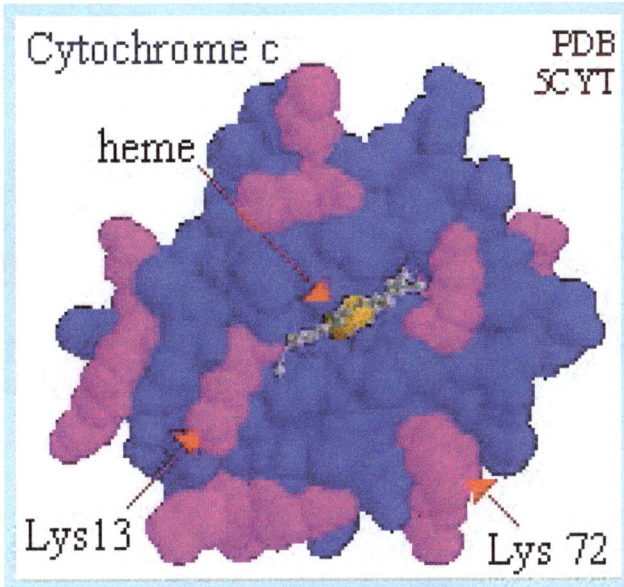

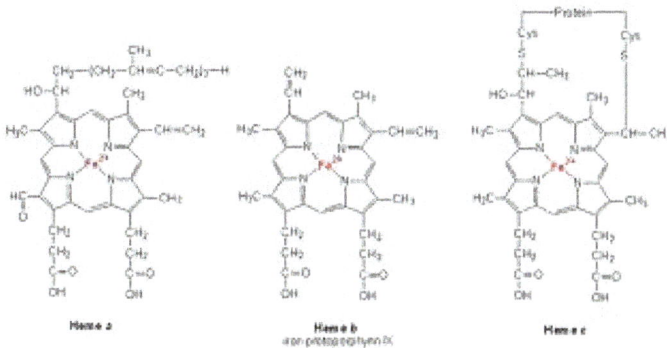

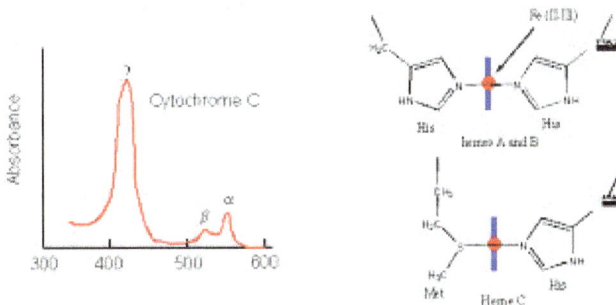

Complex IV (Cytochrome c Oxidase)

Cytochrome c oxidase catalyzes the one electron oxidation of four consecutive four reduced cytochrome in presence of O_2 molecule. The complex is a large integral membrane protein composed of several metal prosthetic sites and 14 protein subunits in mammals. The complex contains two hemes, a cytochrome a and cytochrome a, and two copper centers.

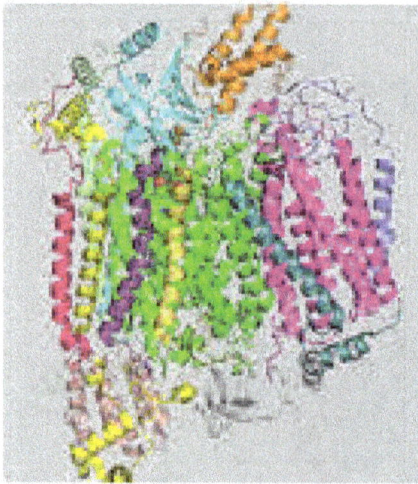

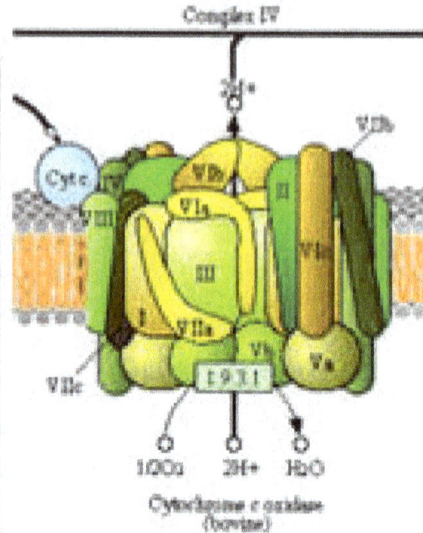

Cytochrome c oxidase
(bovine)

Enzymes

Natural protein that catalyzes chemical reaction taking place within living cell is called enzyme.

Characteristics of Enzyme

- Higher reaction rate

Rate of enzyme catalyzed reaction is often far greater than corresponding reaction catalyzed by synthetic catalyst and 10^6 to 10^{12} higher compared to the corresponding uncatalyzed reaction.

- Milder reaction condition

Enzyme catalyzes reaction in the aqueous medium under very milder condition of temperature, pressure and pH. They work efficiently below 100° C, at atmospheric pressure and at neutral pH.

- Greater reaction specificity

Enzymes are highly specific with respect to the identities of their substrate. Not only that product obtained in an enzyme catalyzed reaction is also specific and there is very rare example where side product also formed.

- Capacity for regulation

The catalytic activities of many enzymes vary in response to the concentrations of many substances other than their substrates. The mechanisms of these regulatory process-es include allosteric control, covalent modification of enzymes, and variation of the amounts of enzymes synthesized.

Simple Enzyme

Enzyme composed of only protein is known as simple enzyme.

Complex Enzyme

Enzyme composed of protein and a relatively a small organic molecule is known as complex enzyme.

Apoenzyme, Cofactor and Holoenzyme

There are many enzymes that comprised of a protein component and a non-protein component. The protein part of the enzyme shows catalytic activity only in the presence of

the non-protein part. Here, the protein part of the enzyme is called apoenzyme and the non- protein part is known as cofactor. The cofactor may be an organic molecule or may be a metal ion. Apoenzyme and cofactor together is known as holoenzyme.

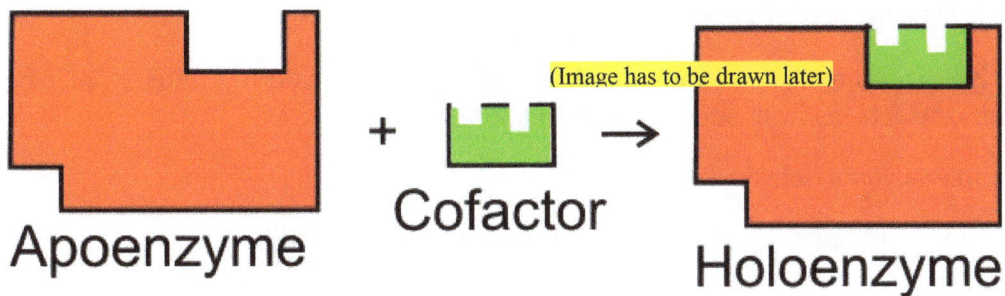

(Image has to be drawn later)

Apoenzyme + Cofactor → Holoenzyme

Coenzyme

When a cofactor is an organic molecule then it is called coenzyme.

Prosthetic Group

When a cofactor is bound so tightly with apoenzyme that it is difficult to remove without damaging the enzyme it is sometimes called prosthetic group.

Classification of Enzymes

(Enzyme Comission-EC)

Based on catalyzed reactions, the Enzyme Commission classified enzyme into six main classes.

Enzymes

Oxidoreductase	Tranferases	Hydrolases	Lyaases	Isomerases	Ligases

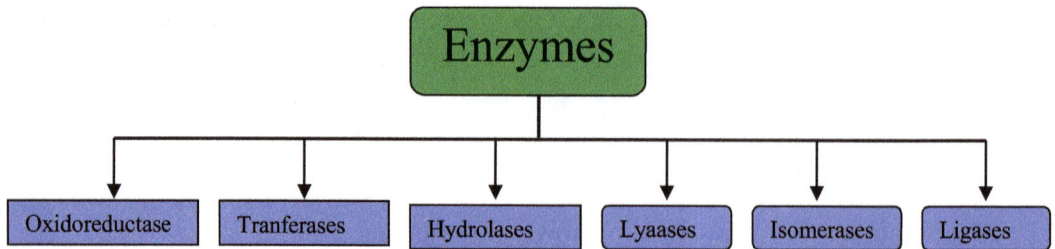

Enzyme class	Type of reaction catalyzed	Example
Oxidoreductase	Oxidation-Reduction reactions	alcohol dehydrogenase
Tranferases	Transfer of a group or atom between two molecules.	phosphorylases
Hydrolases	Hydrolysis reaction	lipase

Lyaases	Removal of a group from substrate (not by hydrolysis)	aldolase
Isomerases	Isomerisation reaction	phosphotriose isom-erase
Ligases	The synthetic joining of two molecules, coupled with the breakdown of pyrophosphate bond in a nucleoside triphosphate.	succinate thiokinase

EC 1. Oxidoreductases

Catalyze a variety of oxidation-reduction reactions by transferring of H atom, oxygen atoms or electrons from one substrate to another.

Example: alcohol dehydrogenase

Reaction: a primary alcohol + NAD$^+$ $\rightleftharpoons$ an aldehyde + NADH + H$^+$

Subclasses

EC 1.1 Acting on the CH-OH group of donors

EC 1.2 Acting on the aldehyde or oxo group of donors

EC 1.3 Acting on the CH-CH group of donors

EC 1.4 Acting on the CH-NH2 group of donors

EC 1.5 Acting on the CH-NH group of donors

EC 1.6 Acting on NADH or NADPH

EC 1.7 Acting on other nitrogenous compounds as donors

EC 1.8 Acting on a sulfur group of donors

EC 1.9Acting on a heme group of donors

EC 1.10 Acting on diphenols and related substances as donors

EC 1.11 Acting on a peroxide as acceptor

EC 1.12 Acting on hydrogen as donor

EC 1.13 Acting on single donors with incorporation of molecular oxygen (oxygenases)

EC 1.14 Acting on paired donors, with incorporation or reduction of molecular oxygen

EC 1.15 Acting on superoxide radicals as acceptor

EC 1.16 Oxidising metal ions

EC 1.17 Acting on CH or CH2 groups

EC 1.18 Acting on iron-sulfur proteins as donors

EC 1.19 Acting on reduced flavodoxin as donor

EC 1.20 Acting on phosphorus or arsenic in donors

EC 1.21 Acting on X-H and Y-H to form an X-Y bond

EC 1.22 Acting on halogen in donors

EC 1.23 Reducing C-O-C group as acceptor

EC 1.97 Other oxidoreductases

EC 2. Transferases

Catalyze those reactions where transfers of groups (acetyl, methyl, phosphate, etc), excluding oxidoreductases (which transfer hydrogen or oxygen and are EC1) occur. These are of the general form:

$$A - X + B \leftrightarrow BX + A$$

Subclasses

EC 2.1 Transferring one-carbon groups

EC 2.2 Transferring aldehyde or ketonic groups

EC 2.3 Acyltransferases

EC 2.4 Glycosyltransferases

EC 2.5 Transferring alkyl or aryl groups, other than methyl groups

EC 2.6 Transferring nitrogenous groups

EC 2.7 Transferring phosphorus-containing groups

EC 2.8 Transferring sulfur-containing groups

EC 2.9 Transferring selenium-containing groups

EC 2.10 Transferring molybdenum- or tungsten-containing groups

EC 3. Hydrolases

Catalyzes the hydrolysis reaction i.e. add water across bond by hydrolyzing it.

EC 3.1 Acting on ester bonds

EC 3.2 Glycosylases

EC 3.3 Acting on ether bonds

EC 3.4 Acting on peptide bonds (peptidases)

EC 3.5 Acting on carbon-nitrogen bonds, other than peptide bonds

EC 3.6 Acting on acid anhydrides

EC 3.7 Acting on carbon-carbon bonds

EC 3.8 Acting on halide bonds

EC 3.9 Acting on phosphorus-nitrogen bonds

EC 3.10 Acting on sulfur-nitrogen bonds

EC 3.11 Acting on carbon-phosphorus bonds

EC 3.12 Acting on sulfur-sulfur bonds

EC 3.13 Acting on carbon-sulfur bonds

EC 4. Lyases

Catalyze the cleavage of C-C, C-O, C-S and C-N bonds by means other than hydrolysis or oxidation.

Common names include decarboxylase and aldolase.

EC 4.1 Carbon-carbon Lyases

Example: pyruvate decarboxylase

Reaction: a 2-oxo carboxylate = an aldehyde + CO_2

EC 4.2 Carbon-oxygen lyases Example: carbonate dehydratase Reaction: H_2CO_3 = CO_2 + H2O

EC 4.3 Carbon-nitrogen lyases Example: aspartate ammonia-lyase Reaction: L-aspartate = fumarate + NH_3

EC 4.4 Carbon-sulfur Lyases

Example: cystathionine γ-lyase

Reaction: L-cystathionine + H_2O = L-cysteine + NH_3 + 2-oxobutanoate

EC 4.5 Carbon-halide Lyases

Example: DDT-dehydrochlorinase

Reaction: 1,1,1-trichloro-2,2-bis(4-chlorophenyl)ethane = 1,1-dichloro-2,2-bis(4- chlorophenyl)ethylene + chloride

EC 4.6 Phosphorus-oxygen Lyases

Example: guanylate cyclase

Reaction: GTP = 3',5'-cyclic GMP + diphosphate

EC 4.7 Carbon-phosphorus Lyases

Example: α-D-ribose 1-methylphosphonate 5-phosphate C-P-lyase

Reaction: α-D-ribose 1-methylphosphonate 5-phosphate = α-D-ribose 1,2-cyclic phosphate 5- phosphate + methane

EC 4.99 Other Lyases

Example: ferrochelatase

Reaction: protoheme + 2 H^+ = protoporphyrin + Fe^{2+}

EC 5. Isomerases

Catalyze atomic rearrangements within a molecule.

EC 5.1 Racemases and Epimerases

Example 1: alanine racemase

Reaction: L-alanine ⇌ D-alanine

Example 2: 4-Hydroxyproline Epimerase

Reaction

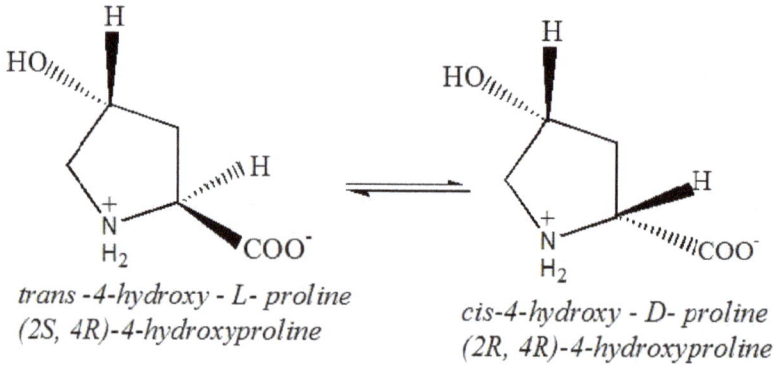

trans -4-hydroxy - L- proline
(2S, 4R)-4-hydroxyproline

cis-4-hydroxy - D- proline
(2R, 4R)-4-hydroxyproline

EC 5.2 *cis-trans*-Isomerases

Example: maleate isomerase

Reaction: maleate ⇌ fumarate

EC 5.3 Intramolecular Oxidoreductases

Example: triose-phosphate isomerase

D-glyceraldehyde 3-phosphate ⇌ glycerone phosphate

EC 5.4 Intramolecular Transferases

Example: lysolecithin acylmutase

Reaction: 2-lysolecithin ⇌ 3-lysolecithin

EC 5.5 Intramolecular Lyases

Example: muconate cycloisomerase

Reaction: 2,5-dihydro-5-oxofuran-2-acetate ⇌ *cis,cis*-hexadienedioate

EC 5.6 Other Isomerases

Example: thiocyanate isomerase

Reaction: benzyl isothiocyanate ⇌ benzyl thiocyanate

EC 6. Ligases

Catalyze the reaction which joins two molecules.

EC 6.1 Forming Carbon—oxygen Bonds

Example: tyrosine—tRNA ligase

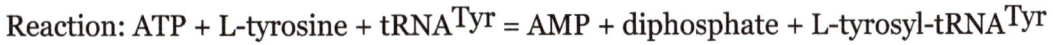

Reaction: ATP + L-tyrosine + $tRNA^{Tyr}$ = AMP + diphosphate + L-tyrosyl-$tRNA^{Tyr}$

EC 6.2 Forming Carbon—sulfur Bonds

Example: acetate—CoA ligase

Reaction: ATP + acetate + CoA = AMP + diphosphate + acetyl-CoA

EC 6.3 Forming Carbon—nitrogen Bonds

Example: aspartate—ammonia ligase

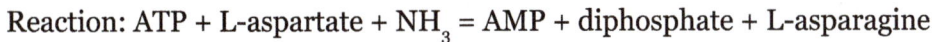

Reaction: ATP + L-aspartate + NH_3 = AMP + diphosphate + L-asparagine

EC 6.4 Forming Carbon—carbon Bonds

Example: pyruvate carboxylase

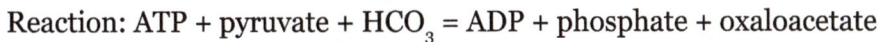

Reaction: ATP + pyruvate + HCO_3 = ADP + phosphate + oxaloacetate

EC 6.5 Forming Phosphoric Ester Bonds

Example: DNA ligase (ATP)

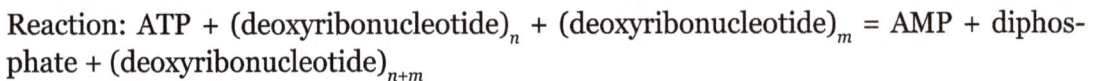

Reaction: ATP + $(deoxyribonucleotide)_n$ + $(deoxyribonucleotide)_m$ = AMP + diphosphate + $(deoxyribonucleotide)_{n+m}$

EC 6.6 Forming Nitrogen—metal Bonds

Example: magnesium chelatase

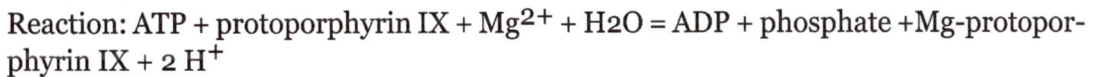

Reaction: ATP + protoporphyrin IX + Mg^{2+} + H2O = ADP + phosphate + Mg-protoporphyrin IX + $2 H^+$

Enzyme

Enzymes are macromolecular biological catalysts. Enzymes accelerate, or catalyze, chemical reactions. The molecules at the beginning of the process upon which enzymes may act are called substrates and the enzyme converts these into different molecules, called products. Almost all metabolic processes in the cell need enzymes in order to occur at rates fast enough to sustain life. The set of enzymes made in a cell determines which metabolic pathways occur in that cell. The study of enzymes is called *enzymology*.

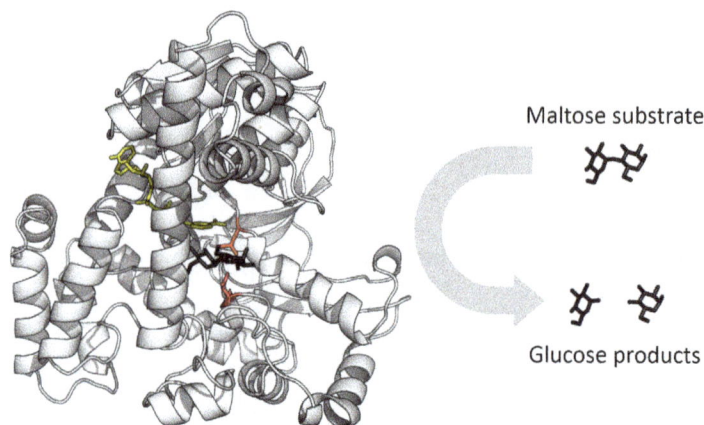

Maltose substrate

Glucose products

The enzyme glucosidase converts the sugar maltose to two glucose sugars. Active site residues in red, maltose substrate in black, and NAD cofactor in yellow. (PDB: 1OBB)

Enzymes are known to catalyze more than 5,000 biochemical reaction types. Most enzymes are proteins, although a few are catalytic RNA molecules. Enzymes' specificity comes from their unique three-dimensional structures.

Like all catalysts, enzymes increase the reaction rate by lowering its activation energy. Some enzymes can make their conversion of substrate to product occur many millions of times faster. An extreme example is orotidine 5'-phosphate decarboxylase, which allows a reaction that would otherwise take millions of years to occur in milliseconds. Chemically, enzymes are like any catalyst and are not consumed in chemical reactions, nor do they alter the equilibrium of a reaction. Enzymes differ from most other catalysts by being much more specific. Enzyme activity can be affected by other molecules: inhibitors are molecules that decrease enzyme activity, and activators are molecules that increase activity. Many drugs and poisons are enzyme inhibitors. An enzyme's activity decreases markedly outside its optimal temperature and pH.

Some enzymes are used commercially, for example, in the synthesis of antibiotics. Some household products use enzymes to speed up chemical reactions: enzymes in biological washing powders break down protein, starch or fat stains on clothes, and enzymes in meat tenderizer break down proteins into smaller molecules, making the meat easier to chew.

Etymology and History

By the late 17th and early 18th centuries, the digestion of meat by stomach secretions and the conversion of starch to sugars by plant extracts and saliva were known but the mechanisms by which these occurred had not been identified.

French chemist Anselme Payen was the first to discover an enzyme, diastase, in 1833. A few decades later, when studying the fermentation of sugar to alcohol by yeast, Louis Pasteur concluded that this fermentation was caused by a vital force contained within the yeast cells called "ferments", which were thought to function only within living organisms. He wrote that "alcoholic fermentation is an act correlated with the life and organization of the yeast cells, not with the death or putrefaction of the cells."

Eduard Buchner

In 1877, German physiologist Wilhelm Kühne (1837–1900) first used the term *enzyme*, which comes from Greek "leavened", to describe this process. The word *enzyme* was used later to refer to nonliving substances such as pepsin, and the word *ferment* was used to refer to chemical activity produced by living organisms.

Eduard Buchner submitted his first paper on the study of yeast extracts in 1897. In a series of experiments at the University of Berlin, he found that sugar was fermented by yeast extracts even when there were no living yeast cells in the mixture. He named the enzyme that brought about the fermentation of sucrose "zymase". In 1907, he received the Nobel Prize in Chemistry for "his discovery of cell-free fermentation". Following Buchner's example, enzymes are usually named according to the reaction they carry out: the suffix -*ase* is combined with the name of the substrate (e.g., lactase is the enzyme that cleaves lactose) or to the type of reaction (e.g., DNA polymerase forms DNA polymers).

The biochemical identity of enzymes was still unknown in the early 1900s. Many scientists observed that enzymatic activity was associated with proteins, but others (such as Nobel laureate Richard Willstätter) argued that proteins were merely carriers for the true enzymes and that proteins *per se* were incapable of catalysis. In 1926, James B. Sumner showed that the enzyme urease was a pure protein and crystallized it; he did likewise for the enzyme catalase in 1937. The conclusion that pure proteins can be enzymes was definitively demonstrated by John Howard Northrop and Wendell Meredith Stanley, who worked on the digestive enzymes pepsin (1930), trypsin and chymotrypsin. These three scientists were awarded the 1946 Nobel Prize in Chemistry.

The discovery that enzymes could be crystallized eventually allowed their structures to be solved by x-ray crystallography. This was first done for lysozyme, an enzyme found in tears, saliva and egg whites that digests the coating of some bacteria; the structure was solved by a group led by David Chilton Phillips and published in 1965. This high-resolution structure of lysozyme marked the beginning of the field of structural biology and the effort to understand how enzymes work at an atomic level of detail.

Naming Conventions

An enzyme's name is often derived from its substrate or the chemical reaction it catalyzes, with the word ending in *-ase*. Examples are lactase, alcohol dehydrogenase and DNA polymerase. Different enzymes that catalyze the same chemical reaction are called isozymes.

The International Union of Biochemistry and Molecular Biology have developed a nomenclature for enzymes, the EC numbers; each enzyme is described by a sequence of four numbers preceded by "EC". The first number broadly classifies the enzyme based on its mechanism.

The top-level classification is:

- EC 1, Oxidoreductases: catalyze oxidation/reduction reactions

- EC 2, Transferases: transfer a functional group (*e.g.* a methyl or phosphate group)

- EC 3, Hydrolases: catalyze the hydrolysis of various bonds

- EC 4, Lyases: cleave various bonds by means other than hydrolysis and oxidation

- EC 5, Isomerases: catalyze isomerization changes within a single molecule

- EC 6, Ligases: join two molecules with covalent bonds.

These sections are subdivided by other features such as the substrate, products, and chemical mechanism. An enzyme is fully specified by four numerical designations. For example, hexokinase (EC 2.7.1.1) is a transferase (EC 2) that adds a phosphate group (EC 2.7) to a hexose sugar, a molecule containing an alcohol group (EC 2.7.1).

Structure

Enzymes are generally globular proteins, acting alone or in larger complexes. Like all proteins, enzymes are linear chains of amino acids that fold to produce a three-dimensional structure. The sequence of the amino acids specifies the structure which in turn determines the catalytic activity of the enzyme. Although structure determines function, a novel enzyme's activity cannot yet be predicted from its structure alone. Enzyme structures unfold (denature) when heated or exposed to chemical denaturants and this disruption to the structure typically causes a loss of activity. Enzyme denaturation is normally linked to temperatures above a species' normal level; as a result, enzymes from bacteria living in volcanic environments such as hot springs are prized by industrial users for their ability to function at high temperatures, allowing enzyme-catalysed reactions to be operated at a very high rate.

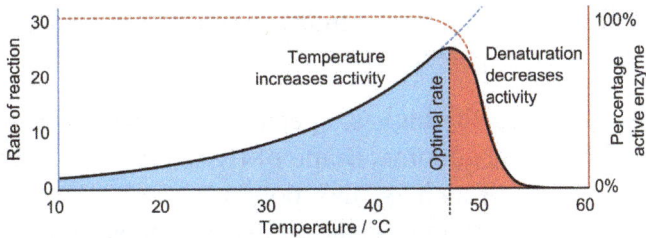

Enzyme activity initially increases with temperature (Q10 coefficient) until the enzyme's structure unfolds (denaturation), leading to an optimal rate of reaction at an intermediate temperature

Enzymes are usually much larger than their substrates. Sizes range from just 62 amino acid residues, for the monomer of 4-oxalocrotonate tautomerase, to over 2,500 residues in the animal fatty acid synthase. Only a small portion of their structure (around 2–4 amino acids) is directly involved in catalysis: the catalytic site. This catalytic site is located next to one or more binding sites where residues orient the substrates. The catalytic site and binding site together comprise the enzyme's active site. The remaining majority of the enzyme structure serves to maintain the precise orientation and dynamics of the active site.

In some enzymes, no amino acids are directly involved in catalysis; instead, the enzyme contains sites to bind and orient catalytic cofactors. Enzyme structures may also contain allosteric sites where the binding of a small molecule causes a conformational change that increases or decreases activity.

A small number of RNA-based biological catalysts called ribozymes exist, which again can act alone or in complex with proteins. The most common of these is the ribosome which is a complex of protein and catalytic RNA components.

Mechanism

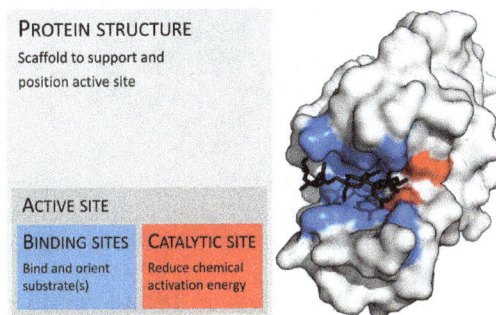

Organisation of enzyme structure and lysozyme example. Binding sites in blue, catalytic site in red and peptidoglycan substrate in black. (PDB: 9LYZ)

Substrate Binding

Enzymes must bind their substrates before they can catalyse any chemical reaction. Enzymes are usually very specific as to what substrates they bind and then the chemical reaction catalysed. Specificity is achieved by binding pockets with complementary shape, charge and hydrophilic/hydrophobic characteristics to the substrates. Enzymes can therefore distinguish between very similar substrate molecules to be chemoselective, regioselective and stereospecific.

Some of the enzymes showing the highest specificity and accuracy are involved in the copying and expression of the genome. Some of these enzymes have "proof-reading" mechanisms. Here, an enzyme such as DNA polymerase catalyzes a reaction in a first step and then checks that the product is correct in a second step. This two-step process results in average error rates of less than 1 error in 100 million reactions in high-fidelity mammalian polymerases. Similar proofreading mechanisms are also found in RNA polymerase, aminoacyl tRNA synthetases and ribosomes.

Conversely, some enzymes display enzyme promiscuity, having broad specificity and acting on a range of different physiologically relevant substrates. Many enzymes possess small side activities which arose fortuitously (i.e. neutrally), which may be the starting point for the evolutionary selection of a new function.

"Lock and key" Model

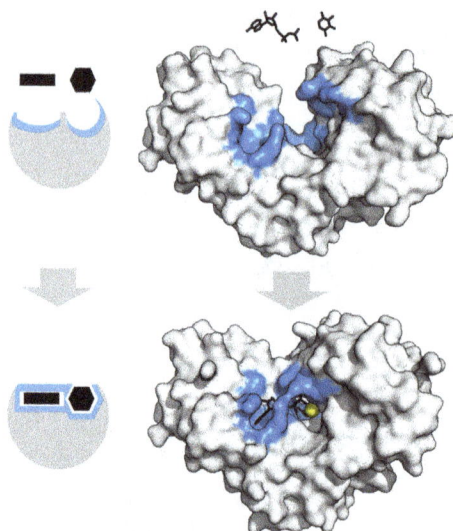

Enzyme changes shape by induced fit upon substrate binding to form enzyme-substrate complex. Hexokinase has a large induced fit motion that closes over the substrates adenosine triphosphate and xylose. Binding sites in blue, substrates in black and Mg^{2+} cofactor in yellow. (PDB: 2E2N, 2E2Q)

To explain the observed specificity of enzymes, in 1894 Emil Fischer proposed that both the enzyme and the substrate possess specific complementary geometric shapes that fit exactly into one another. This is often referred to as "the lock and key" model. This early model explains enzyme specificity, but fails to explain the stabilization of the transition state that enzymes achieve.

Induced Fit Model

In 1958, Daniel Koshland suggested a modification to the lock and key model: since enzymes are rather flexible structures, the active site is continuously reshaped by interactions with the substrate as the substrate interacts with the enzyme. As a result, the substrate does not simply bind to a rigid active site; the amino acid side-chains that make up the active site are molded into the precise positions that enable the enzyme to perform its catalytic function. In some cases, such as glycosidases, the substrate molecule also changes shape slightly as it enters the active site. The active site continues to change until the substrate is completely bound, at which point the final shape and charge distribution is determined. Induced fit may enhance the fidelity of molecular recognition in the presence of competition and noise via the conformational proofreading mechanism.

Catalysis

Enzymes can accelerate reactions in several ways, all of which lower the activation energy ($\Delta G^{\ddagger}$, Gibbs free energy)

1. By stabilizing the transition state:

    o   Creating an environment with a charge distribution complementary to that of the transition state to lower its energy.

2. By providing an alternative reaction pathway:

    o   Temporarily reacting with the substrate, forming a covalent intermediate to provide a lower energy transition state.

3. By destabilising the substrate ground state:

    o   Distorting bound substrate(s) into their transition state form to reduce the energy required to reach the transition state.

    o   By orienting the substrates into a productive arrangement to reduce the reaction entropy change. The contribution of this mechanism to catalysis is relatively small.

Enzymes may use several of these mechanisms simultaneously. For example, proteases such as trypsin perform covalent catalysis using a catalytic triad, stabilise charge build-up on the transition states using an oxyanion hole, complete hydrolysis using an oriented water substrate.

Dynamics

Enzymes are not rigid, static structures; instead they have complex internal dynamic motions – that is, movements of parts of the enzyme's structure such as individual amino acid residues, groups of residues forming a protein loop or unit of secondary structure, or even an entire protein domain. These motions give rise to a conformational ensemble of slightly different structures that interconvert with one another at equilibrium. Different states within this ensemble may be associated with different aspects of an enzyme's function. For example, different conformations of the enzyme dihydrofolate reductase are associated with the substrate binding, catalysis, cofactor release, and product release steps of the catalytic cycle.

Allosteric Modulation

Allosteric sites are pockets on the enzyme, distinct from the active site, that bind to molecules in the cellular environment. These molecules then cause a change in the conformation or dynamics of the enzyme that is transduced to the active site and thus affects the reaction rate of the enzyme. In this way, allosteric interactions can either inhibit or activate enzymes. Allosteric interactions with metabolites upstream or downstream in an enzyme's metabolic pathway cause feedback regulation, altering the activity of the enzyme according to the flux through the rest of the pathway.

Cofactors

Some enzymes do not need additional components to show full activity. Others require non-protein molecules called cofactors to be bound for activity. Cofactors can be either inorganic (e.g., metal ions and iron-sulfur clusters) or organic compounds (e.g., flavin and heme). Organic cofactors can be either coenzymes, which are released from the enzyme's active site during the reaction, or prosthetic groups, which are tightly bound to an enzyme. Organic prosthetic groups can be covalently bound (e.g., biotin in enzymes such as pyruvate carboxylase).

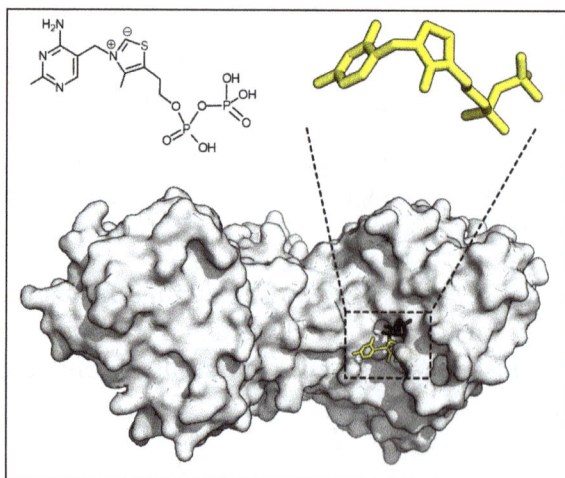

Chemical structure for thiamine pyrophosphate and protein structure of transketolase. Thiamine pyrophosphate cofactor in yellow and xylulose 5-phosphate substrate in black. (PDB: 4KXV)

An example of an enzyme that contains a cofactor is carbonic anhydrase, which is shown in the ribbon diagram above with a zinc cofactor bound as part of its active site. These tightly bound ions or molecules are usually found in the active site and are involved in catalysis. For example, flavin and heme cofactors are often involved in redox reactions.

Enzymes that require a cofactor but do not have one bound are called *apoenzymes* or *apoproteins*. An enzyme together with the cofactor(s) required for activity is called a *holoenzyme* (or haloenzyme). The term *holoenzyme* can also be applied to enzymes that contain multiple protein subunits, such as the DNA polymerases; here the holoenzyme is the complete complex containing all the subunits needed for activity.

Coenzymes

Coenzymes are small organic molecules that can be loosely or tightly bound to an enzyme. Coenzymes transport chemical groups from one enzyme to another. Examples include NADH, NADPH and adenosine triphosphate (ATP). Some coenzymes, such as riboflavin, thiamine and folic acid, are vitamins, or compounds that cannot be synthesized by the body and must be acquired from the diet. The chemical groups carried include the hydride ion (H^-) carried by NAD or $NADP^+$, the phosphate group carried by adenosine triphosphate, the acetyl group carried by coenzyme A, formyl, methenyl or methyl groups carried by folic acid and the methyl group carried by S-adenosylmethionine.

Since coenzymes are chemically changed as a consequence of enzyme action, it is useful to consider coenzymes to be a special class of substrates, or second substrates, which are common to many different enzymes. For example, about 1000 enzymes are known to use the coenzyme NADH.

Coenzymes are usually continuously regenerated and their concentrations maintained at a steady level inside the cell. For example, NADPH is regenerated through the pentose phosphate pathway and *S*-adenosylmethionine by methionine adenosyltransferase. This continuous regeneration means that small amounts of coenzymes can be used very intensively. For example, the human body turns over its own weight in ATP each day.

Thermodynamics

As with all catalysts, enzymes do not alter the position of the chemical equilibrium of the reaction. In the presence of an enzyme, the reaction runs in the same direction as it would without the enzyme, just more quickly. For example, carbonic anhydrase catalyzes its reaction in either direction depending on the concentration of its reactants:

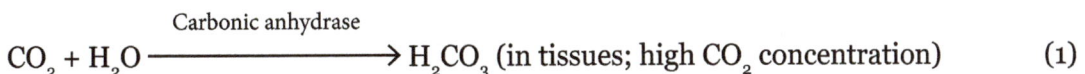

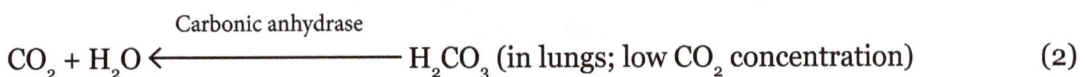

$$CO_2 + H_2O \xrightarrow{\text{Carbonic anhydrase}} H_2CO_3 \text{ (in tissues; high } CO_2 \text{ concentration)} \qquad (1)$$

$$CO_2 + H_2O \xleftarrow{\text{Carbonic anhydrase}} H_2CO_3 \text{ (in lungs; low } CO_2 \text{ concentration)} \qquad (2)$$

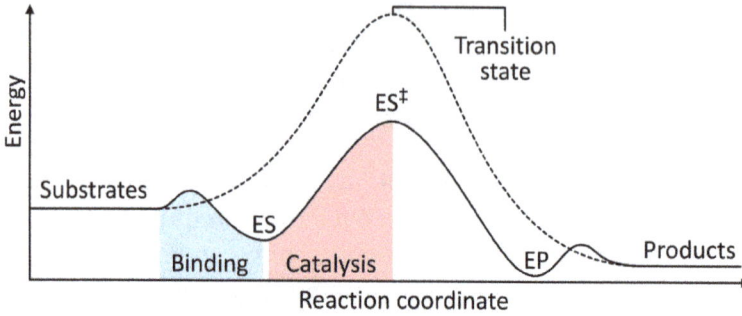

The energies of the stages of a chemical reaction. Uncatalysed (dashed line), substrates need a lot of activation energy to reach a transition state, which then decays into lower-energy products. When enzyme catalysed (solid line), the enzyme binds the substrates (ES), then stabilizes the transition state (ES‡) to reduce the activation energy required to produce products (EP) which are finally released.

The rate of a reaction is dependent on the activation energy needed to form the transition state which then decays into products. Enzymes increase reaction rates by lowering the energy of the transition state. First, binding forms a low energy enzyme-substrate complex (ES). Secondly the enzyme stabilises the transition state such that it requires less energy to achieve compared to the uncatalyzed reaction (ES‡). Finally the enzyme-product complex (EP) dissociates to release the products.

Enzymes can couple two or more reactions, so that a thermodynamically favorable reaction can be used to "drive" a thermodynamically unfavourable one so that the combined energy of the products is lower than the substrates. For example, the hydrolysis of ATP is often used to drive other chemical reactions.

Kinetics

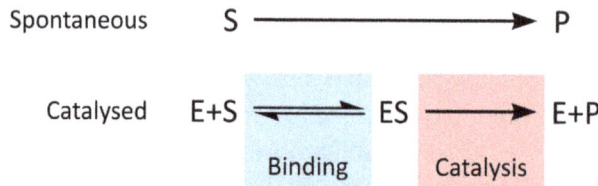

A chemical reaction mechanism with or without enzyme catalysis. The enzyme (E) binds substrate (S) to produce product (P).

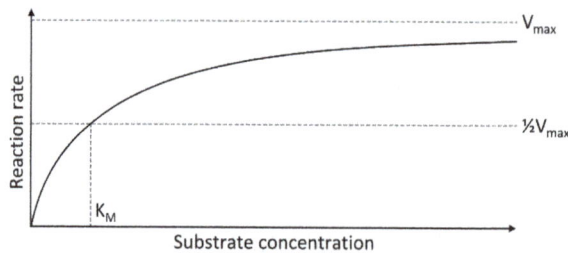

Saturation curve for an enzyme reaction showing the relation between the substrate concentration and reaction rate.

Enzyme kinetics is the investigation of how enzymes bind substrates and turn them into products. The rate data used in kinetic analyses are commonly obtained from enzyme assays. In 1913 Leonor Michaelis and Maud Leonora Menten proposed a quantitative theory of enzyme kinetics, which is referred to as Michaelis–Menten kinetics. The major contribution of Michaelis and Menten was to think of enzyme reactions in two stages. In the first, the substrate binds reversibly to the enzyme, forming the enzyme-substrate complex. This is sometimes called the Michaelis-Menten complex in their honor. The enzyme then catalyzes the chemical step in the reaction and releases the product. This work was further developed by G. E. Briggs and J. B. S. Haldane, who derived kinetic equations that are still widely used today.

Enzyme rates depend on solution conditions and substrate concentration. To find the maximum speed of an enzymatic reaction, the substrate concentration is increased until a constant rate of product formation is seen. This is shown in the saturation curve on the right. Saturation happens because, as substrate concentration increases, more and more of the free enzyme is converted into the substrate-bound ES complex. At the maximum reaction rate (V_{max}) of the enzyme, all the enzyme active sites are bound to substrate, and the amount of ES complex is the same as the total amount of enzyme.

V_{max} is only one of several important kinetic parameters. The amount of substrate needed to achieve a given rate of reaction is also important. This is given by the Michaelis-Menten constant (K_m), which is the substrate concentration required for an enzyme to reach one-half its maximum reaction rate; generally, each enzyme has a characteristic K_m for a given substrate. Another useful constant is k_{cat}, also called the *turnover number*, which is the number of substrate molecules handled by one active site per second.

The efficiency of an enzyme can be expressed in terms of k_{cat}/K_m. This is also called the specificity constant and incorporates the rate constants for all steps in the reaction up to and including the first irreversible step. Because the specificity constant reflects both affinity and catalytic ability, it is useful for comparing different enzymes against each other, or the same enzyme with different substrates. The theoretical maximum for the specificity constant is called the diffusion limit and is about 10^8 to 10^9 ($M^{-1} s^{-1}$). At this point every collision of the enzyme with its substrate will result in catalysis, and the rate of product formation is not limited by the reaction rate but by the diffusion rate. Enzymes with this property are called *catalytically perfect* or *kinetically perfect*. Example of such enzymes are triose-phosphate isomerase, carbonic anhydrase, acetylcholinesterase, catalase, fumarase, β-lactamase, and superoxide dismutase. The turnover of such enzymes can reach several million reactions per second.

Michaelis–Menten kinetics relies on the law of mass action, which is derived from the assumptions of free diffusion and thermodynamically driven random collision. Many biochemical or cellular processes deviate significantly from these conditions, because of macromolecular crowding and constrained molecular movement. More recent, complex extensions of the model attempt to correct for these effects.

Inhibition

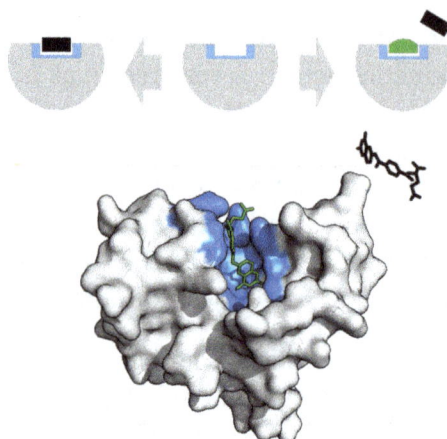

An enzyme binding site that would normally bind substrate can alternatively bind a competitive inhibitor, preventing substrate access. Dihydrofolate reductase is inhibited by methotrexate which prevents binding of its substrate, folic acid. Binding site in blue, inhibitor in green, and substrate in black. (PDB: 4QI9)

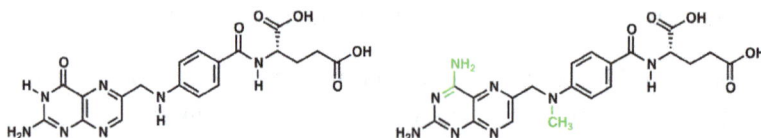

The coenzyme folic acid (left) and the anti-cancer drug methotrexate (right) are very similar in structure (differences show in green). As a result, methotrexate is a competitive inhibitor of many enzymes that use folates.

Enzyme reaction rates can be decreased by various types of enzyme inhibitors.

Types of Inhibition

Competitive

A competitive inhibitor and substrate cannot bind to the enzyme at the same time. Often competitive inhibitors strongly resemble the real substrate of the enzyme. For example, the drug methotrexate is a competitive inhibitor of the enzyme dihydrofolate reductase, which catalyzes the reduction of dihydrofolate to tetrahydrofolate. The similarity between the structures of dihydrofolate and this drug are shown in the accompanying figure. This type of inhibition can be overcome with high substrate concentration. In some cases, the inhibitor can bind to a site other than the binding-site of the usual substrate and exert an allosteric effect to change the shape of the usual binding-site.

Non-competitive

A non-competitive inhibitor binds to a site other than where the substrate binds. The substrate still binds with its usual affinity and hence K_m remains the same. However the inhibitor reduces the catalytic efficiency of the enzyme so that V_{max} is reduced. In contrast to competitive inhibition, non-competitive inhibition cannot be overcome with high substrate concentration.

Uncompetitive

An uncompetitive inhibitor cannot bind to the free enzyme, only to the enzyme-substrate complex; hence, these types of inhibitors are most effective at high substrate concentration. In the presence of the inhibitor, the enzyme-substrate complex is inactive. This type of inhibition is rare.

Mixed

A mixed inhibitor binds to an allosteric site and the binding of the substrate and the inhibitor affect each other. The enzyme's function is reduced but not eliminated when bound to the inhibitor. This type of inhibitor does not follow the Michaelis-Menten equation.

Irreversible

An irreversible inhibitor permanently inactivates the enzyme, usually by forming a covalent bond to the protein. Penicillin and aspirin are common drugs that act in this manner.

Functions of Inhibitors

In many organisms, inhibitors may act as part of a feedback mechanism. If an enzyme produces too much of one substance in the organism, that substance may act as an inhibitor for the enzyme at the beginning of the pathway that produces it, causing production of the substance to slow down or stop when there is sufficient amount. This is a form of negative feedback. Major metabolic pathways such as the citric acid cycle make use of this mechanism.

Since inhibitors modulate the function of enzymes they are often used as drugs. Many such drugs are reversible competitive inhibitors that resemble the enzyme's native substrate, similar to methotrexate above; other well-known examples include statins used to treat high cholesterol, and protease inhibitors used to treat retroviral infections such as HIV. A common example of an irreversible inhibitor that is used as a drug is aspirin, which inhibits the COX_{-1} and COX_{-2} enzymes that produce the inflammation messenger prostaglandin. Other enzyme inhibitors are poisons. For example, the poison cyanide is an irreversible enzyme inhibitor that combines with the copper and iron in the active site of the enzyme cytochrome c oxidase and blocks cellular respiration.

Biological Function

Enzymes serve a wide variety of functions inside living organisms. They are indispensable for signal transduction and cell regulation, often via kinases and phosphatases. They also generate movement, with myosin hydrolyzing ATP to generate muscle contraction, and also transport cargo around the cell as part of the cytoskeleton. Other ATPases in the cell membrane are ion pumps involved in active transport. Enzymes are also involved in more exotic functions, such as luciferase generating light in fireflies. Viruses can also contain enzymes for infecting cells, such as the HIV integrase and reverse transcriptase, or for viral release from cells, like the influenza virus neuraminidase.

An important function of enzymes is in the digestive systems of animals. Enzymes such as amylases and proteases break down large molecules (starch or proteins, respectively) into smaller ones, so they can be absorbed by the intestines. Starch molecules, for example, are too large to be absorbed from the intestine, but enzymes hydrolyze the starch chains into smaller molecules such as maltose and eventually glucose, which can then be absorbed. Different enzymes digest different food substances. In ruminants, which have herbivorous diets, microorganisms in the gut produce another enzyme, cellulase, to break down the cellulose cell walls of plant fiber.

Metabolism

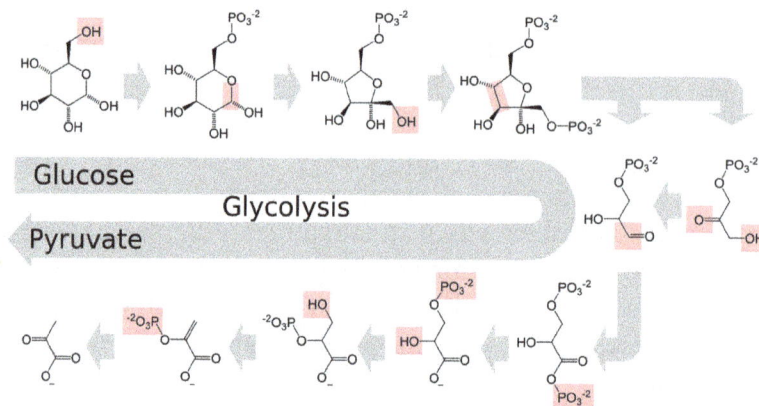

The metabolic pathway of glycolysis releases energy by converting glucose to pyruvate by via a series of intermediate metabolites. Each chemical modification (red box) is performed by a different enzyme.

Several enzymes can work together in a specific order, creating metabolic pathways. In a metabolic pathway, one enzyme takes the product of another enzyme as a substrate. After the catalytic reaction, the product is then passed on to another enzyme. Sometimes more than one enzyme can catalyze the same reaction in parallel; this can allow more complex regulation: with, for example, a low constant activity provided by one enzyme but an inducible high activity from a second enzyme.

Enzymes determine what steps occur in these pathways. Without enzymes, metabolism would neither progress through the same steps and could not be regulated to serve the needs of the cell. Most central metabolic pathways are regulated at a few key steps, typically through enzymes whose activity involves the hydrolysis of ATP. Because this reaction releases so much energy, other reactions that are thermodynamically unfavorable can be coupled to ATP hydrolysis, driving the overall series of linked metabolic reactions.

Control of Activity

There are five main ways that enzyme activity is controlled in the cell.

Regulation

Enzymes can be either activated or inhibited by other molecules. For example, the end product(s) of a metabolic pathway are often inhibitors for one of the first enzymes of the pathway (usually the first irreversible step, called committed step), thus regulating the amount of end product made by the pathways. Such a regulatory mechanism is called a negative feedback mechanism, because the amount of the end product produced is regulated by its own concentration. Negative feedback mechanism can effectively adjust the rate of synthesis of intermediate metabolites according to the demands of the cells. This helps with effective allocations of materials and energy economy, and it prevents the excess manufacture of end products. Like other homeostatic devices, the control of enzymatic action helps to maintain a stable internal environment in living organisms.

Post-translational Modification

Examples of post-translational modification include phosphorylation, myristoylation and glycosylation. For example, in the response to insulin, the phosphorylation of multiple enzymes, including glycogen synthase, helps control the synthesis or degradation of glycogen and allows the cell to respond to changes in blood sugar. Another example of post-translational modification is the cleavage of the polypeptide chain. Chymotrypsin, a digestive protease, is produced in inactive form as chymotrypsinogen in the pancreas and transported in this form to the stomach where it is activated. This stops the enzyme from digesting the pancreas or other tissues before it enters the gut. This type of inactive precursor to an enzyme is known as a zymogen or proenzyme.

Quantity

Enzyme production (transcription and translation of enzyme genes) can be enhanced or diminished by a cell in response to changes in the cell's environment. This form of gene regulation is called enzyme induction. For example, bacteria may become resistant to antibiotics such as penicillin because enzymes called beta-lactamases are induced that hydrolyse the crucial beta-lactam ring within the penicillin molecule. Another example comes from enzymes in the liver called cytochrome P450 oxidases, which are important in drug metabolism. Induction or inhibition of these enzymes can cause drug interactions. Enzyme levels can also be regulated by changing the rate of enzyme degradation.

Subcellular Distribution

Enzymes can be compartmentalized, with different metabolic pathways occurring in different cellular compartments. For example, fatty acids are synthesized by one set of enzymes in the cytosol, endoplasmic reticulum and Golgi and used by a different set of enzymes as a source of energy in the mitochondrion, through β-oxidation. In addition, trafficking of the enzyme to different compartments may change the degree of protonation (cytoplasm neutral and lysosome acidic) or oxidative state [e.g., oxidized (periplasm) or reduced (cytoplasm)] which in turn affects enzyme activity.

Organ Specialization

In multicellular eukaryotes, cells in different organs and tissues have different patterns of gene expression and therefore have different sets of enzymes (known as isozymes) available for metabolic reactions. This provides a mechanism for regulating the overall metabolism of the organism. For example, hexokinase, the first enzyme in the glycolysis pathway, has a specialized form called glucokinase expressed in the liver and pancreas that has a lower affinity for glucose yet is more sensitive to glucose concentration. This enzyme is involved in sensing blood sugar and regulating insulin production.

Involvement in Disease

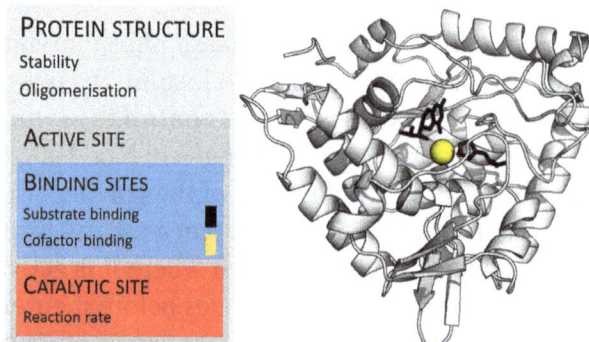

In phenylalanine hydroxylase over 300 different mutations throughout the structure cause phenylketonuria. Phenylalanine substrate and tetrahydrobiopterin coenzyme in black, and Fe^{2+} cofactor in yellow. (PDB: 1KW0)

Since the tight control of enzyme activity is essential for homeostasis, any malfunction (mutation, overproduction, underproduction or deletion) of a single critical enzyme can lead to a genetic disease. The malfunction of just one type of enzyme out of the thousands of types present in the human body can be fatal. An example of a fatal genetic disease due to enzyme insufficiency is Tay-Sachs disease, in which patients lack the enzyme hexosaminidase.

One example of enzyme deficiency is the most common type of phenylketonuria. Many different single amino acid mutations in the enzyme phenylalanine hydroxylase, which catalyzes the first step in the degradation of phenylalanine, result in build-up of phenylalanine and related products. Some mutations are in the active site, directly disrupting binding and catalysis, but many are far from the active site and reduce activity by destabilising the protein structure, or affecting correct oligomerisation. This can lead to intellectual disability if the disease is untreated. Another example is pseudocholinesterase deficiency, in which the body's ability to break down choline ester drugs is impaired. Oral administration of enzymes can be used to treat some functional enzyme deficiencies, such as pancreatic insufficiency and lactose intolerance.

Another way enzyme malfunctions can cause disease comes from germline mutations in genes coding for DNA repair enzymes. Defects in these enzymes cause cancer because cells are less able to repair mutations in their genomes. This causes a slow accumulation of mutations and results in the development of cancers. An example of such a hereditary cancer syndrome is xeroderma pigmentosum, which causes the development of skin cancers in response to even minimal exposure to ultraviolet light.

Industrial Applications

Enzymes are used in the chemical industry and other industrial applications when extremely specific catalysts are required. Enzymes in general are limited in the number of reactions they have evolved to catalyze and also by their lack of stability in organic solvents and at high temperatures. As a consequence, protein engineering is an active area of research and involves attempts to create new enzymes with novel properties, either through rational design or *in vitro* evolution. These efforts have begun to be successful, and a few enzymes have now been designed "from scratch" to catalyze reactions that do not occur in nature.

Enzyme Substrate

When an enzyme unites with its substrate, the resulting union is called enzyme substrate complex.

$$E + S \rightleftharpoons ES$$

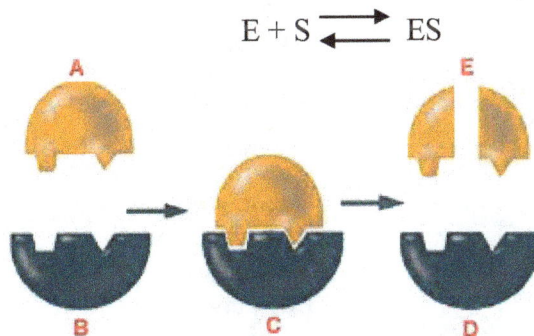

Enzyme Active Site

The active site of an enzyme is a particular location of that enzyme that binds to the substrate and where catalysis occurs. It is constituted with mainly residues of the enzyme and in few cases contains a co-factor. Active site of a particular enzyme possesses a unique geometric shape and chemical properties that allow the enzyme to recognize a specific substrate and bind with it then.

Important properties of active site residues

- Charge (partial, dipoles, helix dipole)
- pKa
- Hydrophobicity
- Flexibility
- Reactivity

Substrate Binding Specificity

One of the most important properties of Enzymes are highly specific with respect to the identity of a substrate. Even, if in a reaction mixture, more than one substrate are present, the enzyme will bind to that substrate only which is complement with the active site with respect to geometric shape, electronic and stereospecificity.

There are two different models to describe how substrate binds with enzyme.

Lock and Key Model

The reaction specificity of an enzyme can be explained with the help of lock and **key** model where enzyme has analogy with lock, active site with key-hole and substrate with key. In case of lock and key, only correctly sized key has permit to open the lock. Similarly, only a substrate which has perfect sized matching with the active site of the enzyme can bind with the enzyme and subsequently reaction can occur.

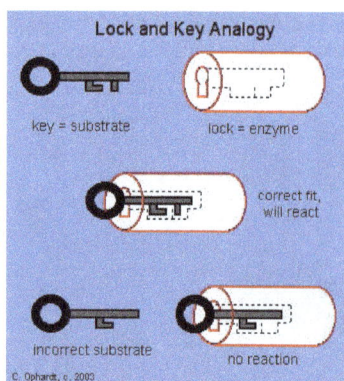

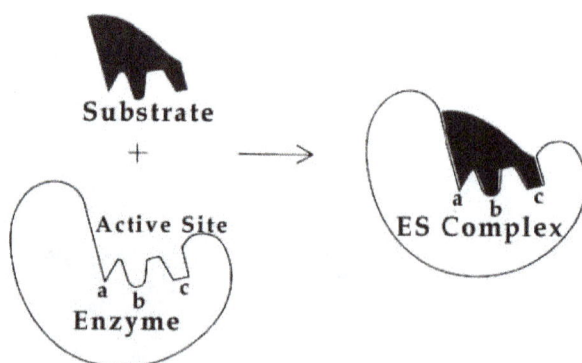

Induced Fit Model

The rigid shaped binding of substrate and enzyme, assumed in the 'lock and key model' can not explain adequately all experimental fact. To find proper explanation induced fit model was proposed. In this model it was assumed that before binding, substrate does not fit exactly with the active site. But after binding, the substrate induce the active site of a structurally flexible enzyme in such a manner that it can fit with it. Thus, with the help of 'induced fit model' it can be explain how a single enzyme is capable of binding many substrate.

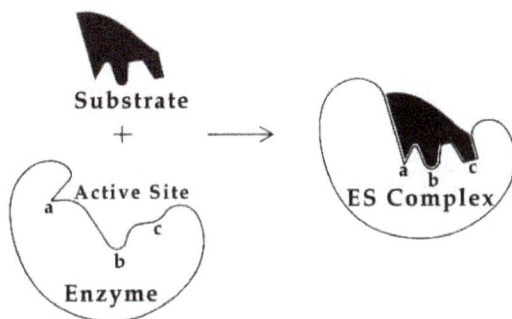

Factors Affecting Enzyme Activity

Enzyme Concentration

Two enzyme molecules bind independently bind with two substrate molecule to give two enzyme substrate complexes whereas one enzyme molecule gives only one. Thus, the rate of an enzymatic reaction increases linearly with increasing enzyme concentration. Deviation from linear relation could be occur

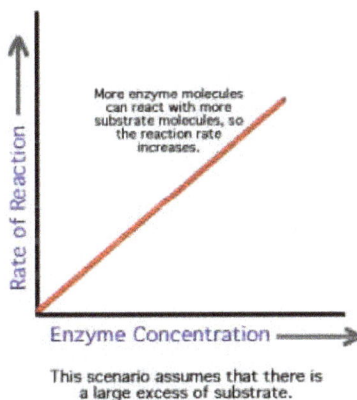

Substrate Concentration

The enzyme catalyzed reaction rate increases as the substrate concentration increases until a certain point called V_{max} is reached where the reaction attains maximal velocity.

Rate of the reaction does not exceed further on increase in substrate concentration because at V_{max} enzyme molecules are completely saturated with substrate molecules. At relatively low substrate concentration the reaction rate increases linearly with increasing substrate concentration. In this case the reaction follows first order kinetics.

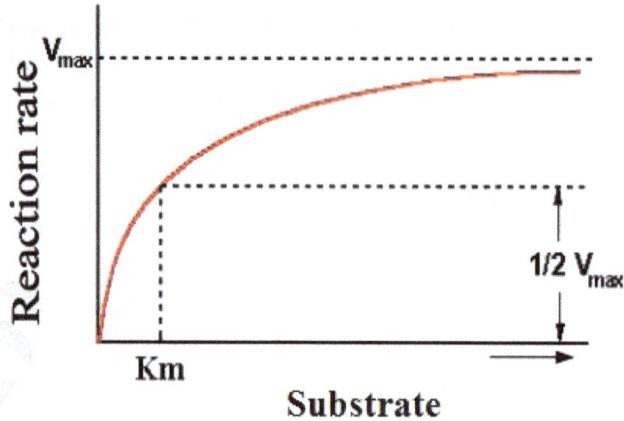

With further increase in substrate concentration the plot becomes curved where rate increase is not as much as in the low substrate concentration. At moderate substrate concentration the reaction follows mixed order kinetics. Once the maximum velocity (V_{max}) point is reached after that with increase in temperature no more velocity is increased. In this case the reaction follows zero order kinetics.

Effect of pH

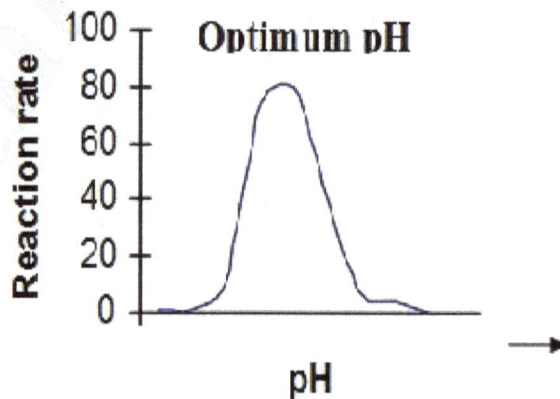

A very little range of pH is effective for an enzyme to be active. Almost for all enzymes there is an optimum temperature where it shows maximum efficiency. The state of optimum condition may be arises due to the following reason: (a) a true reversible effect on its velocity itself, (b) an effect of pH on the affinity of enzyme for the substrate. (c) the effect of pH on the stability of enzyme. Denaturation of the enzyme may takes place on either side of the optimum pH value. All these effects may operate simultaneously.

Effect of Temperature

At very low temperature enzyme does not shows its activity. The rate of an enzymatic reaction increases with increase of temperature until it reaches to the maximum. Further increase of temperature decreases the rate of the reaction. The temperature at which the enzyme attains its maximum velocity is called optimum temperature. The enhancement of reaction velocity with the increase in temperature from low to optimum is due to: (a) With increase in temperature the initial energy the substrate becomes higher which in turn lowers the activation energy and lowers the energy barrier of the reaction. (b) With the rise in temperature the no of collision between enzyme and substrate increases. Decrease in enzyme activity with the rise in temperature beyond optimum temperature is due to denaturation of the enzyme mainly.

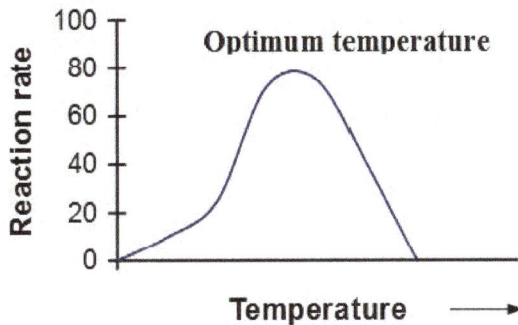

Michaelis- Menten Equation

For an enzyme catalyzed reaction, the plot of intial velocity against substrate concentration gives a hyperbolic curve. The nature of the curve can be explained with the help of Michaelis- Menten equation.

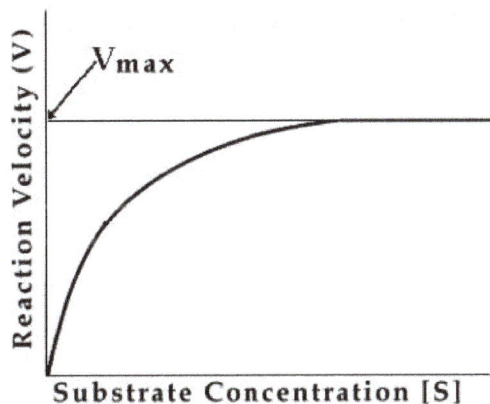

It has been assumed that the overall reaction occurs in two steps. In the first step, enzyme (E) interacts with substrate (S) in a reversible manner to form complex (ES). In the next step, the complex dissociates to give enzyme and product (P) irreversibly. This sequence of event is

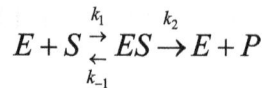

$$E + S \underset{k_{-1}}{\overset{k_1}{\rightleftharpoons}} ES \overset{k_2}{\rightarrow} E + P$$

Now, the initial velocity (rate) of the reaction is

$$v = k_2 [ES] \dots\dots\dots\dots \text{(i)}$$

[ES] can not be measure experimentally. But [ES] is directly related with total enzyme concentration (Et) by the following equation

$$[E_t] = [E] + [ES] \dots\dots\dots\dots \text{(ii)}$$

Rate of ES formation = $k_1 [E][S]$

$$k [E][S] \dots\dots\dots \text{(iii)}$$

$$= k_1 ([E_t] - [ES])[S] \dots\dots\dots \text{(iv)}$$

Rate of ES breakdown

$$= (k_{-1} + k_2)[ES] \dots\dots\dots \text{(v)}$$

Applying the rule of steady state approximation for ES, we have,

Rate of ES formation = Rate of ES breakdown

$$k_1([Et] - [ES])[S] = (k_{-1} + k_2)[ES]$$

$$[ES](k_{-1} + k_2 + k_1[S]) = k_1[E_t][S]$$

$$[ES] = \frac{k_1 [E_t][S]}{k_1 [S] + (k_2 + k_{-1})} \dots\dots\dots \text{(vi)}$$

Dividing the numerator and denominator by k_1

$$[ES] = \frac{[E_t][S]}{[S] + \left(\dfrac{k_2 + k_{-1}}{k_1}\right)} \dots\dots\dots \text{(vii)}$$

Now defining, $\left(\dfrac{k_2 + k_{-1}}{k_1}\right) = K_m$, known as Micheelis- Menten constant, we have,

$$[ES] = \frac{[E_t][S]}{[S] + K_m}$$ .... (viii)

So, the initial velocity (rate) of the reaction is

$$v = k_2[ES]$$ .... (ix)

$$v = \frac{k_2[E_t][S]}{[S] + K_m}$$ .... (x)

At high substrate concentration, all enzyme is saturated with substrate i.e. all enzyme in the ES form and the velocity of the reaction attains its maximum value.

Mathematically,

When $[E_t]$ = [ES], then $v = V$max

Putting these relations in equation (ix) we have,

$$V_{max} = k_2[E_t]$$ .... (xi)

Combining eqn. (x) and (xi), we obtain

$$v = \frac{V_{max}[S]}{[S] + K_m}$$ .... (xii)

This eqn. is known as Michaelis- Menten equation

If the initial velocity, $v = \frac{1}{2}V_{max}$, applying in eqn. (xii) we have

$$\frac{1}{2}V_{max} = \frac{V_{max}[S]}{[S] + K_m}$$

Simplifying, K_m = [S]

Hence, K_m is the substrate concentration when the velocity of the reaction becomes half of its maximum value.

Small K_m value for a enzyme indicates that the reaction acquires maximum catalytic efficiency at low substrate concentration.

Small K_m value indicates tight binding between enzyme and substrate whereas high value of Km indicates weak binding.

Unit of K_m: It has the same unit as that of the substrate concentration.

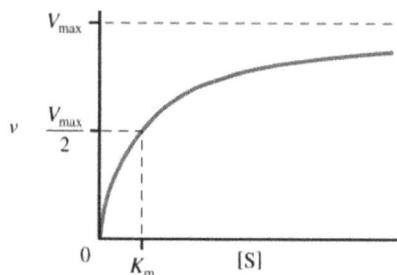

The Lineweaver-Burk plot: *Determination of K_m and V_{max}*

The direct plot of initial velocity (v) vs substrate concentration ([S]) using Micheelis-Menten equation does not give accurate measurement of V_{max} and hence K_m. In this curve, the initial velocity (v) approaches V_{max} asymptotically at very high substrate concentration. But even at 10 time's greater substrate concentration than K_m, the initial velocity (v) is only 91% of that V_{max}. So the value of V_{max}, obtained from the extrapolation of the asymptote will not be accurate. To overcome this problem, Hans Lineweaver and Dean Burk use the reciprocal of eqn. (xii).

$$v = \frac{V_{max}[S]}{[S]+K_m}$$

$$\frac{1}{v} = \frac{[S]+K_m}{V_{max}[S]}$$

$$\frac{1}{v} = \frac{K_m}{V_{max}[S]} + \frac{[S]}{V_{max}[S]}$$

$$\frac{1}{v} = \frac{K_m}{V_{max}}\frac{1}{[S]} + \frac{1}{V_{max}}$$

The plot of 1/v (X- axis) against 1/[S] (Y-axis) generates a straight line with slope = Km/Vmax. This plot is known as Lineweaver- Burk plot or double reciprocal plot. Extrapolation of the straight line gives intercepts 1/Vmax in the Y-axis and -1/Km in the X-axis from which Vmax and Km can be determine easily.

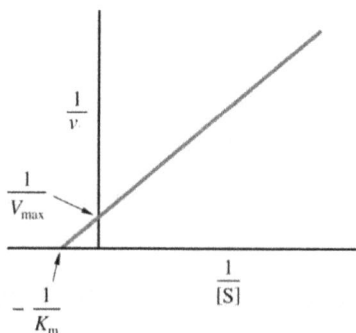

Enzyme Mechanisms

Enzyme binds with the substrate with the help of various interactions like covalent bonding, hydrogen bonding, ionic interactions, ion-dipole and dipole-dipole interactions, charge transfer interactions, hydrophobic interactions, and van der Waals interactions. After binding of the substrate with the active site, the enzyme converts the substrate into the product by employing various types of catalytic mechanisms. These mechanisms are classified as:

- Acid–base catalysis.

- Covalent catalysis.

- Metal ion catalysis.

- Electrostatic catalysis.

- Proximity and orientation effects.

- Preferential binding of the transition state complex.

Acid–Base Catalysis

In general acid catalysis process, lowering of free energy of the transition state occurs via partial proton transfer from a Brønsted acid. For example, keto–enol tautomerization reaction occurs quite slowly when it is uncatalysed. The carabanionlike transition state of this reaction, being high energy slows down the reaction rate. In the presence of acid, donation of proton to the oxygen atom reduces carbanion character of the transition state and hence increases the reaction rate. Few reactions also may be catalyzed by a Brønsted base. Also there are some reactions where simultaneous catalysis by acid and base occur.

The RNase A Reaction Incorporates General Acid–Base Catalysis

An example of enzymatic acid–base catalysis is Hydrolysis of RNA to its component nucleotides by Bovine pancreatic ribonuclease A (RNase A). This digestive enzyme cleaves RNA using the following reaction sequence:

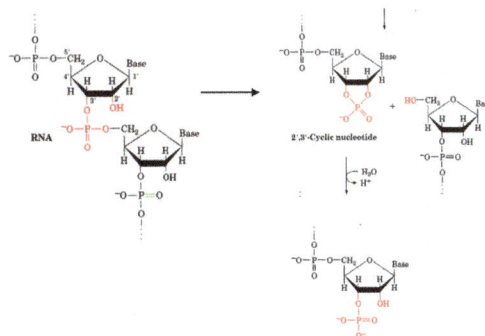

Mechanism

Active site of RNase A comprises of two His residues (His 12 and His 119) and Lys 41 which act in a concerted manner as general acid and base catalysts. RNase A cleaves RNA by following the two-step process:

– In the first step, His 12 behaves as a general base. It abstracts a proton from 2'-OH group of RNA. The newly generated 2'-O⁻, being a good nucleophile promptly attack on the adjacent phosphorous atom. In the concerted way His 119 acts as an acid by protonating the oxygen atom of the leaving group which promotes the bond scission between phosphorous and 5' oxygen. As a result of this step a 2',3'-cyclic intermediate is formed.

– In the second step, actually reverse of the first step take place. Here, His 12 acts as a general acid and His 119 as a general base. His 119 abstract a proton from a water molecule and facilitate the nucleophilic attack to the phosphorous atom of the 2',3'-cyclic intermediate. His 12 protonate the 2'-oxygen atom of the cyclic intermediate and helps to cleave the bond between phosphorous and 2'-oxygen atom.

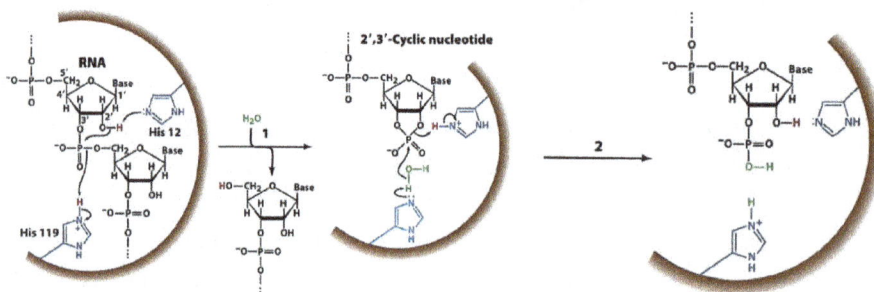

Covalent Catalysis

In the covalent catalysis process a transient catalyst–substrate covalent bond is formed. An example of such a process is decarboxylation of acetoacetate which is chemically catalyzed by primary amines.

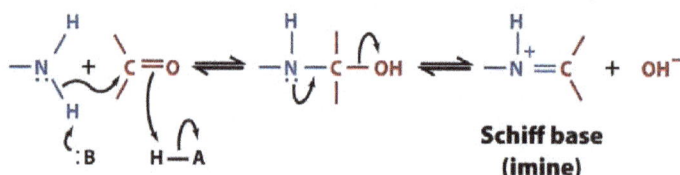

In the uncatalyzed decarboxylation process the enolate transition state is very unstable. Whereas in the catalytic process, the high-energy enolate character of the transition state is stabilized as the protonated nitrogen atom of the covalent intermediate acts as an electron sink. The steps of Schiff base formation as well as its decomposition are very fast, so that these steps are not rate determining in this reaction sequence.

Decarboxylation of acetoacetate

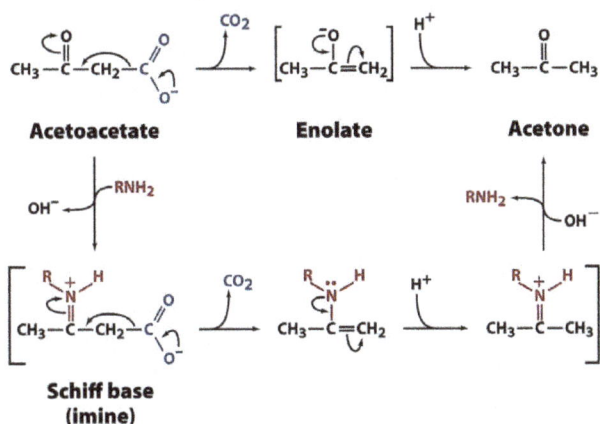

Covalent catalysis is the summation of three successive three stages:

1) The nucleophilic reaction between the catalyst and the substrate to form a covalent bond.

2) The withdrawal of electrons from the reaction center by the now electrophilic catalyst.

3) The elimination of the catalyst, a reaction that is essentially the reverse of stage 1.

D. Electrostatic Catalysis

When substrate binds with the enzyme it generally excludes water from the active site. Due to the presence of substrate at the active site, the local dielectric constant of the active site is similar to that as found in an organic solvent. As a result the electrostatic interactions are much stronger than they are in aqueous solutions. Here, the distribution of charge is present in a medium of low dielectric constant which greatly influences chemical reactivity. Due to the presence of charge group near the active site, the pK's of amino acid side chains in proteins may vary by several units from their nominal values. Distributions of charge near the active sites of enzymes are arranged in such a way that it can stabilize the transition states of the catalyzed reactions which in turn increase the rate of the reaction. Enhancement of enzymatic reaction rate due to the stabilization of transition state by the charge distribution is termed electrostatic catalysis. In case of few enzymatic reactions these charge distributions acts as a guide of a polar substrates to reach toward their binding sites. As a result rate of these reactions is higher than that of their apparent diffusion-controlled limits.

Catalysis Through Proximity and Orientation Effects

Catalytic mechanisms of enzymatic reaction are similar to that of organic model reactions. In spite of this, enzymatic reactions are far more catalytically efficient than these models. Higher efficiency of enzymatic reaction is due to the specific physical conditions at enzyme catalytic sites that promote the corresponding chemical reactions. Proximity and proper orientation of the substrate and active site of an enzyme are the two most obvious specific physical conditions. For a reaction to occur, reactants must come together with the proper spatial relationship.

For example, bimolecular reaction of imidazole with *p*-nitrophenylacetate.

The progress of the reaction is monitored by the formation of the intensely yellow *p*-nitrophenolate ion.

But if the reaction proceeds intramolecularly, the first-order rate constant k2 = 24 k1, when concentration of imidazole is 1M. Thus, due to the attachment of the imidazole catalyst with the the reactant, it is 24-fold more effective than when it is free in solution. So, in the case intramolecular reaction the imidazole group behaves as if its concentration is 24M. Both proximity and orientation contribute to this rate enhancement.

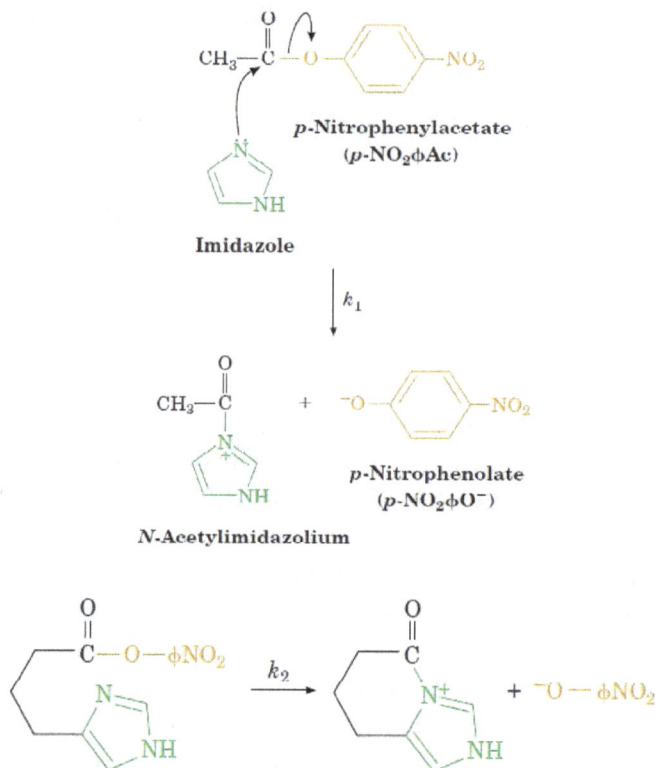

p-Nitrophenylacetate
(p-NO₂φAc)

Imidazole

k_1

p-Nitrophenolate
(p-NO₂φO⁻)

N-Acetylimidazolium

k_2

Catalysis by Preferential Transition State Binding

The catalytic mechanisms so far discussed for an enzyme is not the enough reason for the enormous rate enhancements effected by enzymes. One of the most important mechanisms of enzymatic catalysis is not considered yet i.e., the binding of the transition state to an enzyme with greater affinity than the corresponding substrates or products. If both of them i.e. the previously described catalytic mechanisms and preferential transition state binding are consider together that will rationalizes the observed rates of enzymatic reactions. The concept of transition state binding is based on rack mechanism. The concept proposed that enzymes mechanically strained their substrates toward the transition state geometry through binding sites into which undistorted substrates did not properly fit.

Model (organic reaction) example:

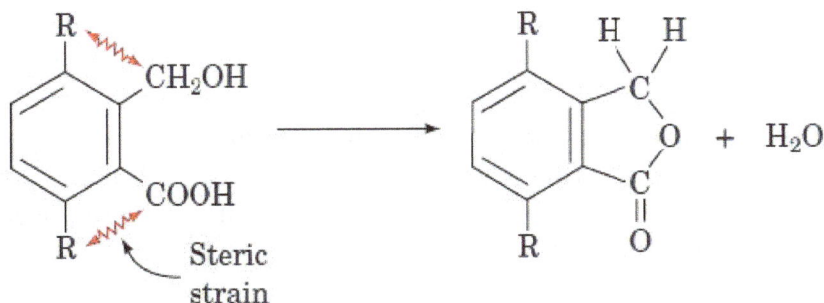

The reaction occurs at 315 times faster rate when R is CH3 rather than when it is H due to the greater steric repulsions between the CH3 groups and the reacting groups. Another example of reaction rate enhancement due to strain is the ring opening reaction where a strained ring such as cyclopropane opens at faster rate compared to the unstained ring such as cyclohexane. In case of both of the mentioned example the strained reactant is more close to the transition state of the reaction in terms of energy than does the corresponding unstrained reactant.

Covalent catalysis are arbitrarily classified as nucleophilic catalysis or electrophilic catalysis. The mechanism of nucleophilic catalysis and general base catalysis are quite similar except that, instead of abstracting a proton from the substrate, the catalyst nucleophilically attacks it so as to form a covalent bond. In covalent catalysis if the covalent bond between catalyst and substrate is more stable the formed, then it will decompose less facilely in the final steps of a reaction. So a good covalent catalyst is one which is a good nucleophile as well as a good leaving group.

Example of Covalent Catalysis:

(1) Hydrolysis of $\beta(1\to4)$ Linkages from N-acetylmuramic acid (NAM) to N-acetylglucosamine (NAG) by Lysozyme:

Lysozyme is an enzyme that destroy bacterial cell wall by hydrolyzing the $\beta(1\to4)$ glycosidic linkages from N-acetylmuramic acid (NAM) to N-acetylglucosamine (NAG) in the alternating NAM–NAG polysaccharide component of cell wall peptidoglycans. It likewise hydrolyzes $\beta(1\to4)$-linked poly(NAG) (chitin), a cell wall component of most fungi.

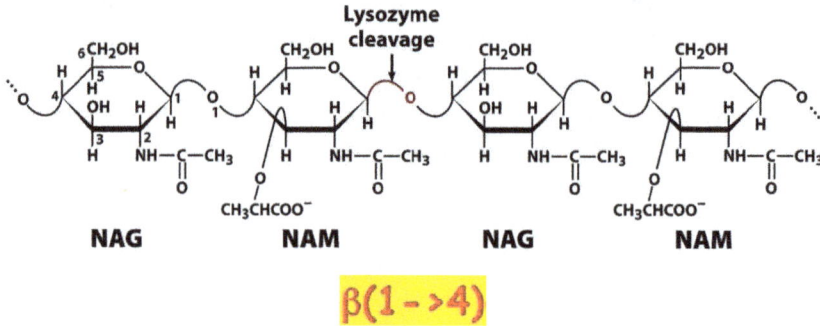

The mechanism of hydrolysis is as follows:

- Lysozyme attaches to a bacterial cell wall by binding to a hexasaccharide unit. In the process, the D-ring is distorted toward the half-chair conformation in response to the unfavorable contacts that its $-C6H2OH$ group would otherwise make with the protein.

- Glu 35 transfers its proton to the O1 atom linking the D- and E-rings, the only polar group in its vicinity, thereby cleaving the C1–O1 bond (general acid catalysis). This step converts the D-ring to a planar resonance-stabilized oxonium ion transition state, whose formation is facilitated by the strain distorting it to the half-chair conformation (catalysis by the preferential binding of the transition state). The positively charged oxonium ion is stabilized by the presence of the nearby negatively charged Asp 52 carboxylate group (electrostatic catalysis). The E-ring product is released.

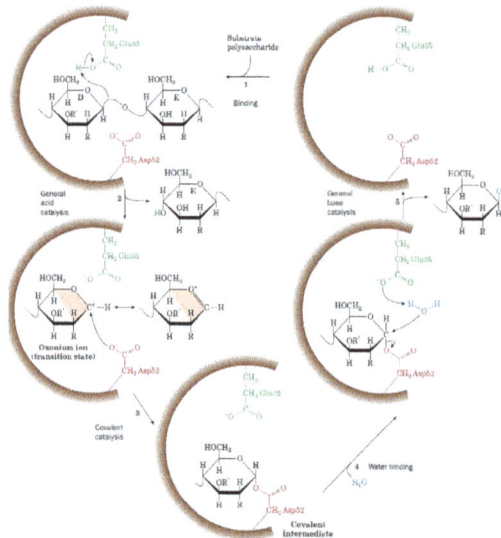

- The Asp 52 carboxylate group nucleophilically attacks the now electron-poor C1 of the D ring to form a covalent glycosyl–enzyme intermediate (covalent catalysis).

- Water replaces the E-ring product in the active site.

- Hydrolysis of the covalent bond with the assistance of Glu 35 (general base catalysis), which involves another oxonium ion transition state, regenerates the active site groups. The enzyme then releases the D-ring product, completing the catalytic cycle.

Enzyme Catalysis

Enzyme catalysis is the increase in the rate of a chemical reaction by the active site of a protein. The protein catalyst (enzyme) may be part of a multi-subunit complex, and/ or may transiently or permanently associate with a Cofactor (e.g. adenosine triphosphate). Catalysis of biochemical reactions in the cell is vital due to the very low reaction rates of the uncatalysed reactions at room temperature and pressure. A key driver of protein evolution is the optimization of such catalytic activities via protein dynamics.

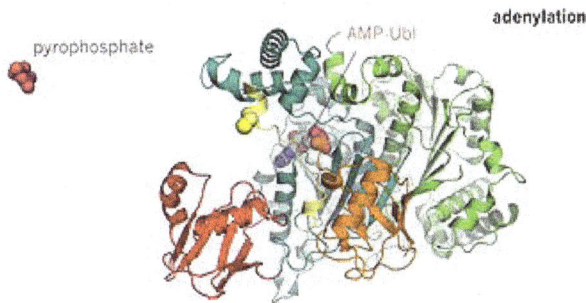

Visualization of ubiquitylation

The mechanism of enzyme catalysis is similar in principle to other types of chemical catalysis. By providing an alternative reaction route the enzyme reduces the energy required to reach the highest energy transition state of the reaction. The reduction of activation energy (Ea) increases the amount of reactant molecules that achieve a sufficient level of energy, such that they reach the activation energy and form the product. As with other catalysts, the enzyme is not consumed during the reaction (as a substrate is) but is recycled such that a single enzyme performs many rounds of catalysis.

Induced Fit

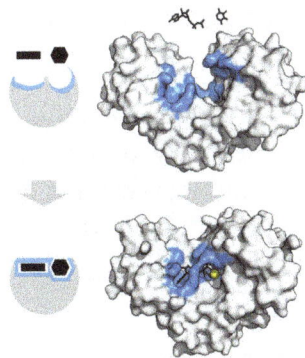

Enzyme changes shape by induced fit upon substrate binding to form enzyme-substrate complex. Hexokinase has a large induced fit motion that closes over the substrates adenosine triphosphate and xylose. Binding sites in blue, substrates in black and Mg²⁺ cofactor in yellow. (PDB: 2E2N, 2E2Q)

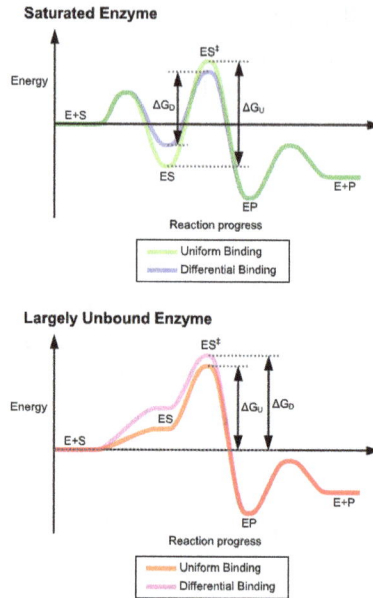

The different mechanisms of substrate binding

The favored model for the enzyme-substrate interaction is the induced fit model. This model proposes that the initial interaction between enzyme and substrate is relatively weak, but that these weak interactions rapidly induce conformational changes in the enzyme that strengthen binding.

The advantages of the induced fit mechanism arise due to the stabilizing effect of strong enzyme binding. There are two different mechanisms of substrate binding: uniform binding, which has strong substrate binding, and differential binding, which has strong transition state binding. The stabilizing effect of uniform binding increases both substrate and transition state binding affinity, while differential binding increases only transition state binding affinity. Both are used by enzymes and have been evolutionarily chosen to minimize the activation energy of the reaction. Enzymes that are saturated, that is, have a high affinity substrate binding, require differential binding to reduce the energy of activation, whereas small substrate unbound enzymes may use either differential or uniform binding.

These effects have led to most proteins using the differential binding mechanism to reduce the energy of activation, so most substrates have high affinity for the enzyme while in the transition state. Differential binding is carried out by the induced fit mechanism - the substrate first binds weakly, then the enzyme changes conformation increasing the affinity to the transition state and stabilizing it, so reducing the activation energy to reach it.

It is important to clarify, however, that the induced fit concept cannot be used to rationalize catalysis. That is, the chemical catalysis is defined as the reduction of $E_a^{\ddagger}$ (when the system is already in the $ES^{\ddagger}$) relative to $E_a^{\ddagger}$ in the uncatalyzed reaction in water (without the enzyme). The induced fit only suggests that the barrier is lower in the closed form of the enzyme but does not tell us what the reason for the barrier reduction is.

Induced fit may be beneficial to the fidelity of molecular recognition in the presence of competition and noise via the conformational proofreading mechanism .

Mechanisms of an Alternative Reaction Route

These conformational changes also bring catalytic residues in the active site close to the chemical bonds in the substrate that will be altered in the reaction. After binding takes place, one or more mechanisms of catalysis lowers the energy of the reaction's transition state, by providing an alternative chemical pathway for the reaction. There are six possible mechanisms of "over the barrier" catalysis as well as a "through the barrier" mechanism:

Proximity and Orientation

The increases the rate of the reaction as enzyme-substrate interactions align reactive chemical groups and hold them close together. This reduces the entropy of the reactants and thus makes reactions such as ligations or addition reactions more favorable, there is a reduction in the overall loss of entropy when two reactants become a single product.

This effect is analogous to an effective increase in concentration of the reagents. The binding of the reagents to the enzyme gives the reaction intramolecular character, which gives a massive rate increase.

For example:

Similar reactions will occur far faster if the reaction is intramolecular.

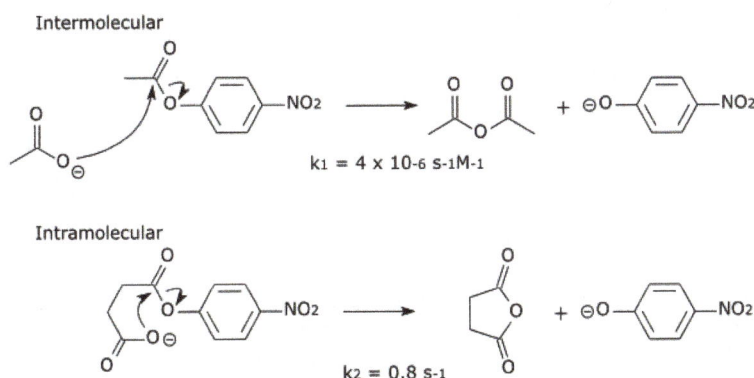

The effective concentration of acetate in the intramolecular reaction can be estimated as $k_2/k_1 = 2 \times 10^5$ Molar

However, the situation might be more complex, since modern computational studies have established that traditional examples of proximity effects cannot be related directly to enzyme entropic effects. Also, the original entropic proposal has been found to largely overestimate the contribution of orientation entropy to catalysis.

Proton Donors or Acceptors

Proton donors and acceptors, i.e. acids and base may donate and accept protons in order to stabilize developing charges in the transition state. This typically has the effect of activating nucleophile and electrophile groups, or stabilizing leaving groups. Histidine is often the residue involved in these acid/base reactions, since it has a pKa close to neutral pH and can therefore both accept and donate protons.

Many reaction mechanisms involving acid/base catalysis assume a substantially altered pKa. This alteration of pKa is possible through the local environment of the residue.

Conditions	Acids	Bases
Hydrophobic environment	Increase pKa	Decrease pKa
Adjacent residues of like charge	Increase pKa	Decrease pKa
Salt bridge (and hydrogen bond) formation	Decrease pKa	Increase pKa

pKa can also be influenced significantly by the surrounding environment, to the extent that residues which are basic in solution may act as proton donors, and vice versa.

For example:

Catalytic Triad of a Serine Protease

The initial step of the serine protease catalytic mechanism involves the histidine of the active site accepting a proton from the serine residue. This prepares the serine as a nucleophile to attack the amide bond of the substrate. This mechanism includes donation of a proton from serine (a base, pKa 14) to histidine (an acid, pKa 6), made possible due to the local environment of the bases.

It is important to clarify that the modification of the pKa's is a pure part of the electrostatic mechanism. Furthermore, the catalytic effect of the above example is mainly associated with the reduction of the pKa of the oxyanion and the increase in the pKa of the histidine, while the proton transfer from the serine to the histidine is not catalyzed significantly, since it is not the rate determining barrier.

Electrostatic Catalysis

Stabilization of charged transition states can also be by residues in the active site forming ionic bonds (or partial ionic charge interactions) with the intermediate. These bonds can either come from acidic or basic side chains found on amino acids such as lysine, arginine, aspartic acid or glutamic acid or come from metal cofactors such as zinc. Metal ions are particularly effective and can reduce the pKa of water enough to make it an effective nucleophile.

Systematic computer simulation studies established that electrostatic effects give, by far, the largest contribution to catalysis. In particular, it has been found that enzyme provides an environment which is more polar than water, and that the ionic transition states are stabilized by fixed dipoles. This is very different from transition state stabilization in water, where the water molecules must pay with "reorganization energy". In order to stabilize ionic and charged states. Thus, the catalysis is associated with the fact that the enzyme polar groups are preorganized.

The magnitude of the electrostatic field exerted by an enzyme's active site has been shown to be highly correlated with the enzyme's catalytic rate enhancement

Binding of substrate usually excludes water from the active site, thereby lowering the local dielectric constant to that of an organic solvent. This strengthens the electrostatic interactions between the charged/polar substrates and the active sites. In addition, studies have shown that the charge distributions about the active sites are arranged so as to stabilize the transition states of the catalyzed reactions. In several enzymes, these charge distributions apparently serve to guide polar substrates toward their binding sites so that the rates of these enzymatic reactions are greater than their apparent diffusion-controlled limits.

For example:

Carboxypeptidase Catalytic Mechanism

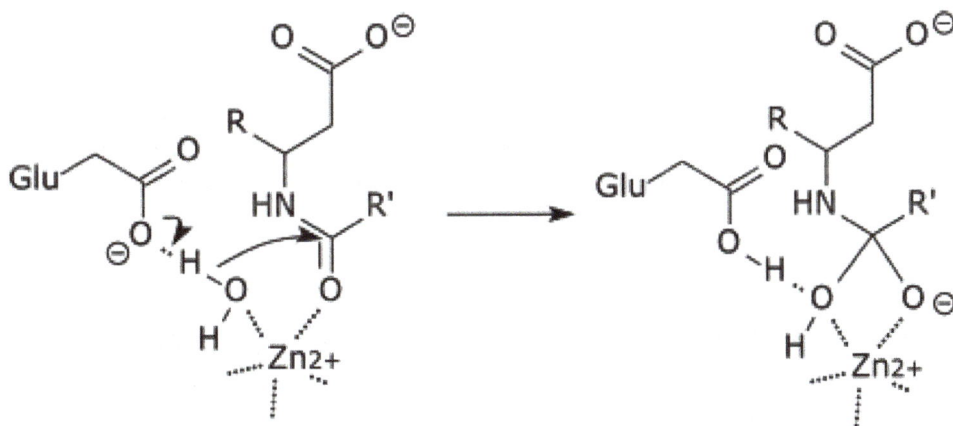

The tetrahedral intermediate is stabilised by a partial ionic bond between the Zn^{2+} ion and the negative charge on the oxygen

Covalent Catalysis

Covalent catalysis involves the substrate forming a transient covalent bond with residues in the enzyme active site or with a cofactor. This adds an additional covalent intermediate to the reaction, and helps to reduce the energy of later transition states of the reaction. The covalent bond must, at a later stage in the reaction, be broken to regenerate the enzyme. This mechanism is utilised by the catalytic triad of enzymes such as proteases like chymotrypsin and trypsin, where an acyl-enzyme intermediate is formed. An alternative mechanism is schiff base formation using the free amine from a lysine residue, as seen in the enzyme aldolase during glycolysis.

Some enzymes utilize non-amino acid cofactors such as pyridoxal phosphate (PLP) or thiamine pyrophosphate (TPP) to form covalent intermediates with reactant molecules. Such covalent intermediates function to reduce the energy of later transition states, similar to how covalent intermediates formed with active site amino acid residues allow stabilization, but the capabilities of cofactors allow enzymes to carryout reactions that amino acid side residues alone could not. Enzymes utilizing such cofactors include the PLP-dependent enzyme aspartate transaminase and the TPP-dependent enzyme pyruvate dehydrogenase.

Rather than lowering the activation energy for a reaction pathway, covalent catalysis provides an alternative pathway for the reaction (via to the covalent intermediate) and so is distinct from true catalysis. For example, the energetics of the covalent bond to the serine molecule in chymotrypsin should be compared to the well-understood covalent bond to the nucleophile in the uncatalyzed solution reaction. A true proposal of a covalent catalysis (where the barrier is lower than the corresponding barrier in solution) would require, for example, a partial covalent bond to the transition state by an enzyme group (e.g., a very strong hydrogen bond), and such effects do not contribute significantly to catalysis.

Metal Ion Catalysis

The presence of a metal ion in the active site participates in catalysis by coordinating charge stabilization and shielding. Because of a metal's positive charge, only negative charges can be stabilized through metal ions. Metal ions can also act to ionize water by acting as a Lewis acid. Metal ions may also be agents of oxidation and reduction.

Bond Strain

This is the principal effect of induced fit binding, where the affinity of the enzyme to the transition state is greater than to the substrate itself. This induces structural rearrangements which strain substrate bonds into a position closer to the conformation of the transition state, so lowering the energy difference between the substrate and transition state and helping catalyze the reaction.

However, the strain effect is, in fact, a ground state destabilization effect, rather than transition state stabilization effect. Furthermore, enzymes are very flexible and they cannot apply large strain effect.

In addition to bond strain in the substrate, bond strain may also be induced within the enzyme itself to activate residues in the active site.

For example:

Substrate, bound substrate, and transition state conformations of lysozyme.

The substrate, on binding, is distorted from the half chair conformation of the hexose ring (because of the steric hindrance with amino acids of the protein forcing the equatorial c6 to be in the axial position) into the chair conformation.

Quantum Tunneling

These traditional "over the barrier" mechanisms have been challenged in some cases by models and observations of "through the barrier" mechanisms (quantum tunneling). Some enzymes operate with kinetics which are faster than what would be predicted by the classical $\Delta G^{\ddagger}$. In "through the barrier" models, a proton or an electron can tunnel through activation barriers. Quantum tunneling for protons has been observed in tryptamine oxidation by aromatic amine dehydrogenase.

Interestingly, quantum tunneling does not appear to provide a major catalytic advantage, since the tunneling contributions are similar in the catalyzed and the uncatalyzed reactions in solution. However, the tunneling contribution (typically enhancing rate constants by a factor of ~1000 compared to the rate of reaction for the classical 'over the barrier' route) is likely crucial to the viability of biological organisms. This emphasizes the general importance of tunneling reactions in biology.

In 1971-1972 the first quantum-mechanical model of enzyme catalysis was formulated.

Active Enzyme

The binding energy of the enzyme-substrate complex cannot be considered as an external energy which is necessary for the substrate activation. The enzyme of high energy content may firstly transfer some specific energetic group X_1 from catalytic site of the enzyme to the final place of the first bound reactant, then another group X_2 from the second bound reactant (or from the second group of the single reactant) must be transferred to active site to finish substrate conversion to product and enzyme regeneration.

We can present the whole enzymatic reaction as a two coupling reactions:

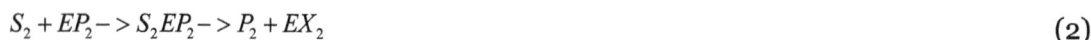

$$S_1 + EX_1 -> S_1EX_1 -> P_1 + EP_2 \tag{1}$$

$$S_2 + EP_2 -> S_2EP_2 -> P_2 + EX_2 \tag{2}$$

It may be seen from reaction (1) that the group X_1 of the active enzyme appears in the product due to possibility of the exchange reaction inside enzyme to avoid both electrostatic inhibition and repulsion of atoms. So we represent the active enzyme as a powerful reactant of the enzymatic reaction. The reaction (2) shows incomplete conversion of the substrate because its group X_2 remains inside enzyme. This approach as idea had formerly proposed relying on the hypothetical extremely high enzymatic conversions (catalytically perfect enzyme).

The crucial point for the verification of the present approach is that the catalyst must be a complex of the enzyme with the transfer group of the reaction. This chemical aspect is supported by the well-studied mechanisms of the several enzymatic reactions. Let us consider the reaction of peptide bond hydrolysis catalyzed by a pure protein α-chymotrypsin (an enzyme acting without a cofactor), which is a well-studied member of the serine proteases family.

We present the experimental results for this reaction as two chemical steps:

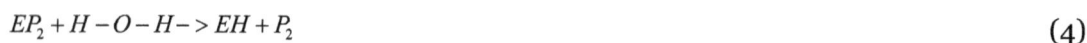

$$S_1 + EH -> P_1 + EP_2 \tag{3}$$

$$EP_2 + H-O-H -> EH + P_2 \tag{4}$$

where S_1 is a polypeptide, P_1 and P_2 are products. The first chemical step (3) includes the formation of a covalent acyl-enzyme intermediate. The second step (4) is the deacylation step. It is important to note that the group H+, initially found on the enzyme, but not in water, appears in the product before the step of hydrolysis, therefore it may be considered as an additional group of the enzymatic reaction.

Thus, the reaction (3) shows that the enzyme acts as a powerful reactant of the reaction. According to the proposed concept, the H transport from the enzyme promotes the first reactant conversion, breakdown of the first initial chemical bond (between groups P_1 and P_2). The step of hydrolysis leads to a breakdown of the second chemical bond and regeneration of the enzyme.

The proposed chemical mechanism does not depend on the concentration of the substrates or products in the medium. However, a shift in their concentration mainly causes free energy changes in the first and final steps of the reactions (1) and (2) due to the changes in the free energy content of every molecule, whether S or P, in water solution.

This approach is in accordance with the following mechanism of muscle contraction. The final step of ATP hydrolysis in skeletal muscle is the product release caused by the association of myosin heads with actin. The closing of the actin-binding cleft during the association reaction is structurally coupled with the opening of the nucleotide-binding pocket on the myosin active site.

Notably, the final steps of ATP hydrolysis include the fast release of phosphate and the slow release of ADP. The release of a phosphate anion from bound ADP anion into water solution may be considered as an exergonic reaction because the phosphate anion has low molecular mass.

Thus, we arrive at the conclusion that the primary release of the inorganic phosphate $H_2PO_4^-$ leads to transformation of a significant part of the free energy of ATP hydrolysis into the kinetic energy of the solvated phosphate, producing active streaming. This assumption of a local mechano-chemical transduction is in accord with Tirosh's mechanism of muscle contraction, where the muscle force derives from an integrated action of active streaming created by ATP hydrolysis.

Examples of Catalytic Mechanisms

In reality, most enzyme mechanisms involve a combination of several different types of catalysis.

Triose Phosphate Isomerase

Triose phosphate isomerase (EC 5.3.1.1) catalyses the reversible interconvertion of the two triose phosphates isomers dihydroxyacetone phosphate and D-glyceraldehyde 3-phosphate.

Trypsin

Trypsin (EC 3.4.21.4) is a serine protease that cleaves protein substrates after lysine or arginine residues using a catalytic triad to perform covalent catalysis, and an oxyanion hole to stabilise charge-buildup on the transition states.

Aldolase

Aldolase (EC 4.1.2.13) catalyses the breakdown of fructose 1,6-bisphosphate (F-1,6-BP) into glyceraldehyde 3-phosphate and dihydroxyacetone phosphate (DHAP).

Enzyme Diffusivity

The advent of single-molecule studies led in the 2010s to the observation that the movement of untethered enzymes increases with increasing substrate concentration and increasing reaction enthalpy. Subsequent observations suggest that this increase in diffusivity is driven by transient displacement of the enzyme's center of mass, resulting in a "recoil effect that propels the enzyme".

Reaction Similarity

Similarity between enzymatic reactions (EC) can be calculated by using bond changes, reaction centres or substructure metrics (EC-BLAST).

Serine Proteases

Another example of covalent catalysis is a diverse group of proteolytic enzymes known as the serine proteases. Their name 'serine proteases' comes from the presence of a reactive Ser residue in their active site and this residue takes part in catalytic activity following a common mechanism. We will discuss the mechanism of chymotrypsin (a serine protease) mediated proteolysis.

First chymotrypsin binds to the substrate to form the Michaelis complex. Then in the rate- determining step of the reaction, Ser 195 nucleophilically attacks to the carbonyl group of scissile peptide to form a complex known as the tetrahedral intermediate (covalent catalysis). His 57 abstract the liberated proton to form an imidazolium ion (general base catalysis).

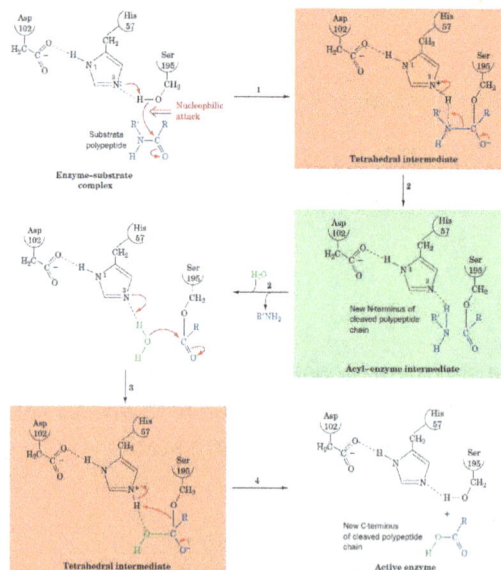

102 is hydrogen bonded to His 57. The polarizing effect of the unsolvated carboxylate ion of Asp 102 helps His 57 to abstract the proton

In the second step, N3 of His 57 donates a proton to the N atom of tetrahedral intermediate. It provides driving force for the decomposition of tetrahedral intermediate into the acyl–enzyme intermediate. The leaving group (R'NH2) of the reaction is the new N- terminal portion of the cleaved polypeptide chain. It is released from the enzyme and replaced by water from the solvent.

In the third step, deacylation of the acyl–enzyme intermediate takes place rapidly. Water nucleophilically attacks to the acyl carbon of the acyl–enzyme intermediate which is facilitating by the abstraction of proton by His 57 from that water molecule. The process is followed by the leaving of Ser 195 which in turn results carboxylate product (the new C- terminal portion of the cleaved polypeptide chain).

Metal Ion Catalysis

Presence of metal ions is quite essential for many enzymes to show catalytic activity. Depending on the strengths of their ion–protein interactions metal ion catalyst can be classified into two subclasses:

–   Metalloenzymes: Here, the protein part of the enzyme is tightly bound with metal ions. These metal ions are generally transition metal ion, such as Fe^{2+}, Fe^{3+}, Cu^{2+}, Zn^{2+}, Mn^{2+}, or Co^{3+}.

–   Metal-activated enzymes: Here, the protein part of the enzyme is loosely bound with metal ions from solution. These metal ions are usually the alkali and alkaline earth metal ions, such as Na^+, K^+, Mg^{2+}, or Ca^{2+}.

There are three major ways by which metal ions participate in the catalytic process in:

1) By binding to substrates so as to orient them properly for reaction.

2) By mediating oxidation–reduction reactions through reversible changes in the metal ion's oxidation state.

3) By electrostatically stabilizing or shielding negative charges.

Among the three major ways, the third aspect of metal ion catalysis will be discussed here.

Metal Ions Promote Catalysis through Charge Stabilization

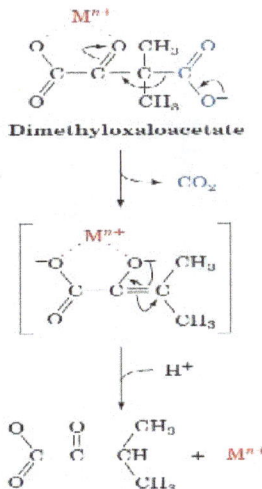

Dimethyloxaloacetate

Similar to proton, metal ion acts by neutralizing negative charge in many metal ion–catalyzed reactions. Here, the metal ions behave as a Lewis acid. As a catalyst Metal ion has advantage over proton as metal ions can be present in high concentrations at neutral pH's can have charges greater than +1. So, Metal ions are more effective catalysts compared to protons.

An example nonenzymatic catalysis by a metal ion is the decarboxylation of dimethyloxaloacetate catalyzed by metal ions such as Cu^{2+} and Ni^{2+}.

Metal Ions Promote Nucleophilic Catalysis via Water Ionization

When a metal ion binds with water molecules, the charge of the metal ion causes that water molecule to be more acidic than free H_2O and therefore a source of OH^- ions even below neutral pH's. For example, the water molecule of $(NH_3)_5Co^{3+}(H_2O)$ ionizes according to the reaction:

$$(NH_3)_5Co^{3+}(H_2O) \leftrightarrow (NH_3)_5Co^{3+}(OH^-) + H^+$$

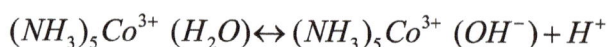

Carbonic Anhydrase

An example of mechanism of metal ion catalysis is catalytic mechanism of the following reaction by carbonic anhydrase:

$$CO_2 + H_2O \leftrightarrow HCO_3^- + H+$$

In carbonic anhydrase a Zn^{2+} ion is present at the active site cleft. The central metal atom is tetrahedrally coordinated by three imidazole (Im) ring of three His side chains and an O atom of either an HCO_3^- ion or a water molecule. The enzyme acts by the sequence of following catalytic mechanism:

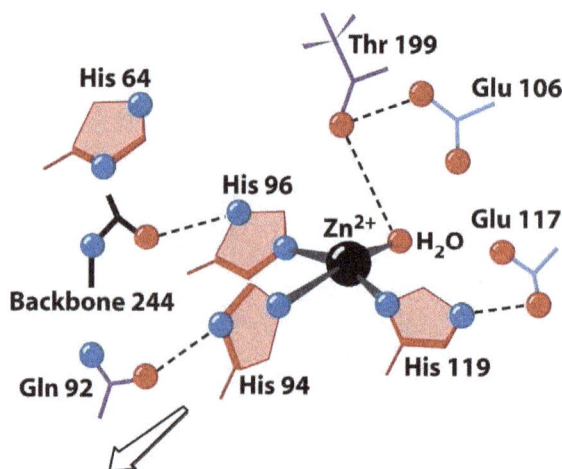

At first, the metal ion i.e. Zn^{2+} polarizes H_2O molecule. This water molecule then ionizes in a process facilitated through general base catalysis by His 64. Although His 64 is too far away from the Zn^{2+}-bound water to directly abstract its proton, these entities are linked by two intervening water molecules to form a hydrogen bonded network that is thought to act as a proton shuttle.

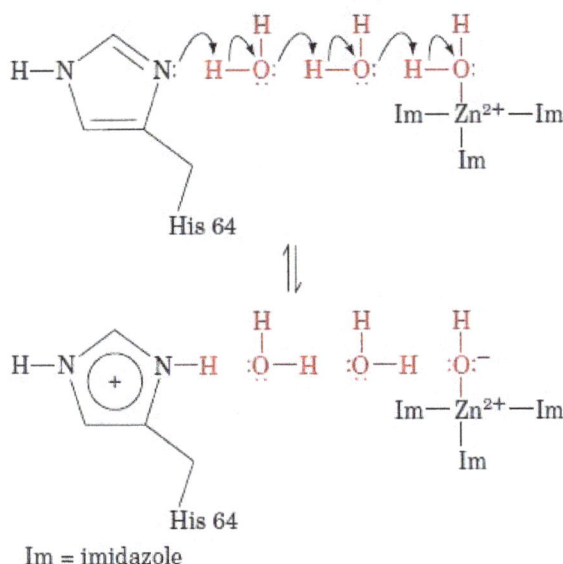

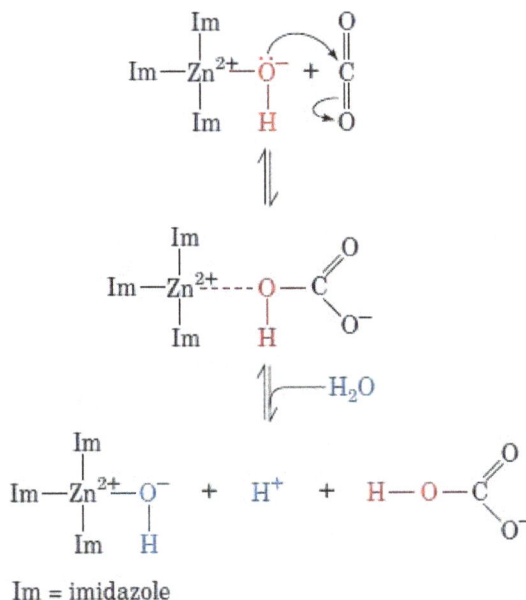

His 64

His 64

Im = imidazole

Im = imidazole

The OH^- ion resulting from the former step, nucleophilically attacks the nearby enzymatically bound CO_2, thereby converting it to HCO_3^-. The optimal geometry of OH^- for nucleophilic attack on the substrate CO_2 is attaining by the existence of two hydrogen bond. The first hydrogen bond is between Zn^{2+}-bound OH^- group with Thr 199 and the second one is between Thr with Glu 106.

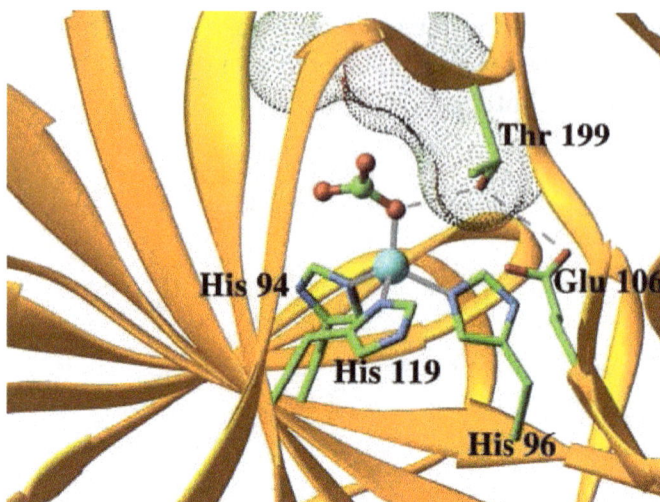

In the final step, the catalytic site is regenerated. Here, water molecule completely replaces HCO_3^- from the Zn^{2+}-bound HCO_3^- and deprotonation of His 64 also occur.

Michaelis–Menten Kinetics

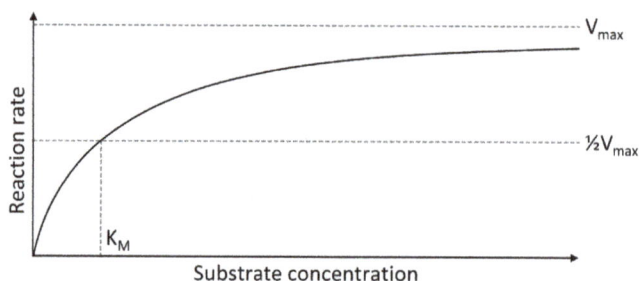

Michaelis–Menten saturation curve for an enzyme reaction showing the relation between the substrate concentration and reaction rate

In biochemistry, *'Michaelis–Menten' kinetics* is one of the best-known models of enzyme kinetics. It is named after German biochemist Leonor Michaelis and Canadian physician Maud Menten. The model takes the form of an equation describing the rate of enzymatic reactions, by relating reaction rate v to $[S]$, the concentration of a substrate S. Its formula is given by

$$v = \frac{d[P]}{dt} = \frac{V_{max}[S]}{K_M + [S]}.$$

This equation is called the Michaelis–Menten equation. Here, V_{max} represents the maximum rate achieved by the system, at saturating substrate concentration. The Michaelis constant K_M is the substrate concentration at which the reaction rate is half of V_{max}. Biochemical reactions involving a single substrate are often assumed to follow Michaelis–Menten kinetics, without regard to the model's underlying assumptions.

Model

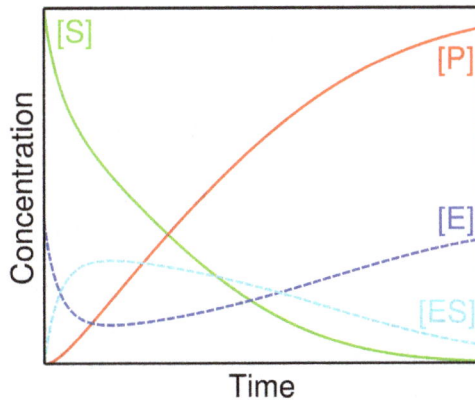

Change in concentrations over time for enzyme E, substrate S, complex ES and product P

In 1903, French physical chemist Victor Henri found that enzyme reactions were initiated by a bond (more generally, a binding interaction) between the enzyme and the substrate. His work was taken up by German biochemist Leonor Michaelis and Canadian physician Maud Menten, who investigated the kinetics of an enzymatic reaction mechanism, invertase, that catalyzes the hydrolysis of sucrose into glucose and fructose. In 1913, they proposed a mathematical model of the reaction. It involves an enzyme, E, binding to a substrate, S, to form a complex, ES, which in turn releases a product, P, regenerating the original enzyme. This may be represented schematically as

$$E + S \underset{k_r}{\overset{k_f}{\rightleftarrows}} ES \xrightarrow{\ k_{cat}\ } E + p$$

where k_f (forward rate), k_r (reverse rate), and k_{cat} (catalytic rate) denote the rate constants, the double arrows between S (substrate) and ES (enzyme-substrate complex) represent the fact that enzyme-substrate binding is a reversible process, and the single forward arrow represents the formation of P (product).

Under certain assumptions – such as the enzyme concentration being much less than the substrate concentration – the rate of product formation is given by

$$v = \frac{d[P]}{dt} = V_{max} \frac{[S]}{K_M + [S]} = k_{cat}[E]_0 \frac{[S]}{K_M + [S]}.$$

The reaction order depends on the relative size of the two terms in the denominator.

At low substrate concentration $[S] \ll K_M$ so that $v = k_{cat}[E]_0 \frac{[S]}{K_M}$. Under these conditions the reaction rate varies linearly with substrate concentration [S] (first-order kinetics). However at higher [S] with $[S] \gg K_M$, the reaction becomes independent of [S] (zero-order kinetics) and asymptotically approaches its maximum rate $V_{max} = k_{cat}[E]_0$, where $[E]_0$ is the initial enzyme concentration. This rate is attained when all enzyme is bound to substrate. k_{cat}, the turnover number, is the maximum number of substrate molecules converted to product per enzyme molecule per second. Further addition of substrate does not increase the rate which is said to be saturated.

The Michaelis constant K_M is the [S] at which the reaction rate is at half-maximum, and is an inverse measure of the substrate's affinity for the enzyme—as a small K_M indicates high affinity, meaning that the rate will approach V_{max} with lower [S] than those reactions with a larger K_M. The value of K_M is dependent on both the enzyme and the substrate, as well as conditions such as temperature and pH.

The model is used in a variety of biochemical situations other than enzyme-substrate interaction, including antigen-antibody binding, DNA-DNA hybridization, and protein-protein interaction. It can be used to characterise a generic biochemical reaction, in the same way that the Langmuir equation can be used to model generic adsorption of biomolecular species. When an empirical equation of this form is applied to microbial growth, it is sometimes called a Monod equation.

Applications

Parameter values vary widely between enzymes:

Enzyme	K_M (M)	k_{cat} (1/s)	k_{cat} / K_M (1/(M*s))
Chymotrypsin	1.5×10^{-2}	0.14	9.3
Pepsin	3.0×10^{-4}	0.50	1.7×10^3
Tyrosyl-tRNA synthetase	9.0×10^{-4}	7.6	8.4×10^3
Ribonuclease	7.9×10^{-3}	7.9×10^2	1.0×10^5
Carbonic anhydrase	2.6×10^{-2}	4.0×10^5	1.5×10^7
Fumarase	5.0×10^{-6}	8.0×10^2	1.6×10^8

The constant k_{cat} / K_M (catalytic efficiency) is a measure of how efficiently an enzyme converts a substrate into product. Diffusion limited enzymes, such as fumarase, work at the theoretical upper limit of $10^8 - 10^{10}$ /M*s, limited by diffusion of substrate into the active site.

Michaelis–Menten kinetics have also been applied to a variety of spheres outside of biochemical reactions, including alveolar clearance of dusts, the richness of species pools, clearance of blood alcohol, the photosynthesis-irradiance relationship, and bacterial phage infection.

Derivation

Applying the law of mass action, which states that the rate of a reaction is proportional to the product of the concentrations of the reactants (i.e. $[E][S]$), gives a system of four non-linear ordinary differential equations that define the rate of change of reactants with time t

$$\frac{d[E]}{dt} = -k_f[E][S] + k_r[ES] + k_{cat}[ES]$$

$$\frac{d[S]}{dt} = -k_f[E][S] + k_r[ES]$$

$$\frac{d[ES]}{dt} = k_f[E][S] - k_r[ES] - k_{cat}[ES]$$

$$\frac{d[P]}{dt} = k_{cat}[ES].$$

In this mechanism, the enzyme E is a catalyst, which only facilitates the reaction, so that its total concentration, free plus combined, $[E]+[ES]=[E]_0$ is a constant. This conservation law can also be observed by adding the first and third equations above.

Equilibrium Approximation

In their original analysis, Michaelis and Menten assumed that the substrate is in instantaneous chemical equilibrium with the complex, which implies

$$k_f[E][S] = k_r[ES].$$

From the enzyme conservation law, we obtain

$$[E] = [E]_0 - [ES].$$

Combining the two expressions above, gives us

$$k_f([E]_0 - [ES])[S] = k_r[ES].$$

Upon simplification, we get

$$[ES] = \frac{[E]_0[S]}{K_d + [S]}$$

where $K_d = k_r / k_f$ is the dissociation constant for the enzyme-substrate complex. Hence the velocity v of the reaction – the rate at which P is formed – is

$$v = \frac{d[P]}{dt} = \frac{V_{max}[S]}{K_d + [S]}$$

where $V_{max} = k_{cat}[E]_0$ is the maximum reaction velocity.

Quasi-steady-state Approximation

An alternative analysis of the system was undertaken by British botanist G. E. Briggs and British geneticist J. B. S. Haldane in 1925. They assumed that the concentration of the intermediate complex does not change on the time-scale of product formation – known as the quasi-steady-state assumption or pseudo-steady-state-hypothesis. Mathematically, this assumption means $k_f[E][S] = k_r[ES] + k_{cat}[ES]$. Combining this relationship with the enzyme conservation law, the concentration of the complex is

$$[ES] = \frac{[E]_0[S]}{K_M + [S]}$$

where

$$K_M = \frac{k_r + k_{cat}}{k_f}$$

is known as the Michaelis constant, where k_r, k_{cat}, and k_f are, respectively, the constants for substrate unbinding, conversion to product, and binding to the enzyme. Hence the velocity v of the reaction is

$$v = \frac{d[P]}{dt} = \frac{V_{max}[S]}{K_M + [S]}.$$

Assumptions and Limitations

The first step in the derivation applies the law of mass action, which is reliant on free diffusion. However, in the environment of a living cell where there is a high concentration of proteins, the cytoplasm often behaves more like a gel than a liquid, limiting molecular movements and altering reaction rates. Although the law of mass action can be valid in heterogeneous environments, it is more appropriate to model the cytoplasm as a fractal, in order to capture its limited-mobility kinetics.

The resulting reaction rates predicted by the two approaches are similar, with the only difference being that the equilibrium approximation defines the constant as K_d, whilst the quasi-steady-state approximation uses K_M. However, each approach is founded upon a different assumption. The Michaelis–Menten equilibrium analysis is valid if the substrate reaches equilibrium on a much faster time-scale than the product is formed or, more precisely, that

$$\epsilon_d = \frac{k_{cat}}{k_r} \ll 1.$$

By contrast, the Briggs–Haldane quasi-steady-state analysis is valid if

$$\epsilon_m = \frac{[E]_0}{[S]_0 + K_M} \ll 1.$$

Thus it holds if the enzyme concentration is much less than the substrate concentration. Even if this is not satisfied, the approximation is valid if K_M is large.

In both the Michaelis–Menten and Briggs–Haldane analyses, the quality of the approximation improves as $\dot{o}$ decreases. However, in model building, Michaelis–Menten kinetics are often invoked without regard to the underlying assumptions.

It is also important to remember that, while irreversibility is a necessary simplification in order to yield a tractable analytic solution, in the general case product formation is not in fact irreversible. The enzyme reaction is more correctly described as

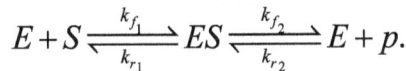

$$E + S \underset{k_{r_1}}{\overset{k_{f_1}}{\rightleftharpoons}} ES \underset{k_{r_2}}{\overset{k_{f_2}}{\rightleftharpoons}} E + p.$$

In general, the assumption of irreversibility is a good one in situations where one of the below is true:

1. The concentration of substrate(s) is very much larger than the concentration of products:

$$[S] \gg [P].$$

This is true under standard *in vitro* assay conditions, and is true for many *in vivo* biological reactions, particularly where the product is continually removed by a subsequent reaction.

2. The energy released in the reaction is very large, that is

$$\Delta G \ll 0.$$

In situations where neither of these two conditions hold (that is, the reaction is low energy and a substantial pool of product(s) exists), the Michaelis–Menten equation breaks down, and more complex modelling approaches explicitly taking the forward and reverse reactions into account must be taken to understand the enzyme biology.

Determination of Constants

The typical method for determining the constants V_{max} and K_M involves running a series of enzyme assays at varying substrate concentrations $[S]$, and measuring the initial reaction rate v_0. 'Initial' here is taken to mean that the reaction rate is measured after a relatively short time period, during which it is assumed that the enzyme-substrate complex has formed, but that the substrate concentration held approximately constant, and so the equilibrium or quasi-steady-state approximation remain valid. By plotting reaction rate against concentration, and using nonlinear regression of the Michaelis–Menten equation, the parameters may be obtained.

Before computing facilities to perform nonlinear regression became available, graphical methods involving linearisation of the equation were used. A number of these were proposed, including the Eadie–Hofstee diagram, Hanes–Woolf plot and Lineweaver–Burk plot; of these, the Hanes–Woolf plot is the most accurate. However, while useful for visualization, all three methods distort the error structure of the data and are inferior to nonlinear regression. Nonetheless, their use can still be found in modern literature.

In 1997 Santiago Schnell and Claudio Mendoza suggested a closed form solution for the time course kinetics analysis of the Michaelis–Menten kinetics based on the solution of the Lambert W function. Namely:

$$\frac{[S]}{K_M} = W(F(t))$$

where W is the Lambert W function and

$$F(t) = \frac{[S]_0}{K_M} \exp\left(\frac{[S]_0}{K_M} - \frac{V_{max}}{K_M} t \right).$$

The above equation has been used to estimate V_{max} and K_M from time course data.

Role of Substrate Unbinding

The Michaelis-Menten equation has been used to predict the rate of product formation in enzymatic reactions for more than a century. Specifically, it states that the rate of an enzymatic reaction will increase as substrate concentration increases, and that increased unbinding of enzyme-substrate complexes will decrease the reaction rate. While the first prediction is well established, the second has never been tested experimentally. To determine whether an increased rate of unbinding does in fact decrease the reaction rate, Shlomi Reuveni *et al.* mathematically analyzed the effect of enzyme-substrate unbinding on enzymatic reactions at the single-molecule level. According to the study, unbinding of an enzyme from a substrate can reduce the rate of product formation under some conditions, but may also have the opposite effect. As substrate concentrations increase, a tipping point can be reached where an increase in the unbinding rate results in an increase, rather than a decrease, of the reaction rate. The results indicate that enzymatic reactions can behave in ways that violate the classical Michaelis-Menten equation, and that the role of unbinding in enzymatic catalysis still remains to be determined experimentally.

References

- Murphy JM, et al. (2014). "A robust methodology to subclassify pseudokinases based on their nucleotide-binding properties". Biochemical Journal. 457 (2): 323–334. PMID 24107129. doi:10.1042/BJ20131174

- Petsko GA, Ringe D (2003). "Chapter 1: From sequence to structure". Protein structure and function. London: New Science. p. 27. ISBN 978-1405119221

- de Bolster M (1997). "Glossary of Terms Used in Bioinorganic Chemistry: Cofactor". International Union of Pure and Applied Chemistry. Retrieved 30 October 2007

- Smith S (December 1994). "The animal fatty acid synthase: one gene, one polypeptide, seven enzymes". FASEB Journal. 8 (15): 1248–59. PMID 8001737

- Suzuki H (2015). "Chapter 7: Active Site Structure". How Enzymes Work: From Structure to Function. Boca Raton, FL: CRC Press. pp. 117–140. ISBN 978-981-4463-92-8

- Radzicka A, Wolfenden R (January 1995). "A proficient enzyme". Science. 267 (5194): 90–931. Bibcode:1995Sci...267...90R. PMID 7809611. doi:10.1126/science.7809611

- Briggs GE, Haldane JB (1925). "A Note on the Kinetics of Enzyme Action". The Biochemical Journal. 19 (2): 339–339. PMC 1259181. PMID 16743508. doi:10.1042/bj0190338

- Krauss G (2003). "The Regulations of Enzyme Activity". Biochemistry of Signal Transduction and Regulation (3rd ed.). Weinheim: Wiley-VCH. pp. 89–114. ISBN 9783527605767

- "BRENDA The Comprehensive Enzyme Information System". Technische Universität Braunschweig. Retrieved 23 February 2015

- Mackie RI, White BA (October 1990). "Recent advances in rumen microbial ecology and metabolism: potential impact on nutrient output". Journal of Dairy Science. 73 (10): 2971–95. PMID 2178174. doi:10.3168/jds.S0022-0302(90)78986-2

- Cooper GM (2000). "Chapter 2.2: The Central Role of Enzymes as Biological Catalysts". The Cell: a Molecular Approach (2nd ed.). Washington (DC): ASM Press. ISBN 0-87893-106-6

- Callahan BP, Miller BG (December 2007). "OMP decarboxylase—An enigma persists". Bioorganic Chemistry. 35 (6): 465–9. PMID 17889251. doi:10.1016/j.bioorg.2007.07.004

- Faergeman NJ, Knudsen J (April 1997). "Role of long-chain fatty acyl-CoA esters in the regulation of metabolism and in cell signalling". The Biochemical Journal. 323 (Pt 1): 1–12. PMC 1218279. PMID 9173866

- Cox MM, Nelson DL (2013). "Chapter 6.2: How enzymes work". Lehninger Principles of Biochemistry (6th ed.). New York, N.Y.: W.H. Freeman. p. 195. ISBN 978-1464109621

- Manchester KL (December 1995). "Louis Pasteur (1822–1895)–chance and the prepared mind". Trends in Biotechnology. 13 (12): 511–5. PMID 8595136. doi:10.1016/S0167-7799(00)89014-9

- Schnell, S.; Mendoza, C. (1997). "A closed form solution for time-dependent enzyme kinetics". Journal of Theoretical Biology. 187 (2): 207–212. doi:10.1006/jtbi.1997.0425

- Voet, Donald; Voet, Judith; Pratt, Charlotte (2016). Fundamentals of Biochemistry. Hoboken, New Jersey: John Wiley & Sons, Inc. p. 336. ISBN 978-1-118-91840-1

- Anfinsen CB (July 1973). "Principles that govern the folding of protein chains". Science. 181 (4096): 223–30. Bibcode:1973Sci...181..223A. PMID 4124164. doi:10.1126/science.181.4096.223

PERMISSIONS

Index